Introduction to Embedded Systems and Robotics

Nayan M. Kakoty • Rupam Goswami
Ramana Vinjamuri

Introduction to Embedded Systems and Robotics

A Practical Guide

Nayan M. Kakoty
School of Engineering
Tezpur University
Tezpur, Assam, India

Rupam Goswami
School of Engineering
Tezpur University
Tezpur, Assam, India

Ramana Vinjamuri
University of Maryland Baltimore County
Ellicott City, NJ, USA

ISBN 978-3-031-73100-6   ISBN 978-3-031-73098-6   (eBook)
https://doi.org/10.1007/978-3-031-73098-6

This Springer imprint is published by the registered company Springer Nature Switzerland AG
The registered company address is: Gewerbestrasse 11, 6330 Cham, Switzerland

If disposing of this product, please recycle the paper.

Preface

The study and development of robots can be traced back to the sketches in the 1495's notebook of Leonardo da Vinci. The first industrial robot by UNIMATION was introduced in 1950 for operations such as lifting the pieces of metal from die-casting machines. For obvious advantages of using robots, its applications expanded to manufacturing industry, entertainment, healthcare technology, extra-terrestrial space exploration, coaching in games, and military applications since the beginning of twenty-first century. Gradual increase of robot applications has necessitated them to be more user-friendly, and the days are not far when assistive robotics will be part of everyone's life during daily living activities.

For easy understanding of the core domains of engineering and technology involve in robotics, this book has been prepared systematically to introduce the fundamental concepts of embedded systems and robotics. The embedded systems, an integral part of a robot, act like the brain of it which process the information and command to control to robot activities. An embedded system acquires the information about the robot's surroundings through sensory inputs and command the actuators to perform motions or movements accordingly.

The readers will be able to grasp the physics and technical information necessary to start robot development while reading the chapters. This will be facilitated through the lens of *ABC* for robotics. In brief, the *ABC* pedagogy, i.e., *Acquire*, *Build*, and *Create*, of the book aims to garner interest in the readers through practical examples. The readers are expected to first *Acquire* the basic information about embedded systems and robotics. Thereafter, they shall be equipped with knowledge to *Build* different circuits, systems, and sub-systems based on the acquired information. Finally, they shall be ready to *Create* simple solutions to practical problems with the gained concepts.

This book is devised to train the reader realizing the textbook concepts, which can bring changes in a physical world. An embedded system is one of the easier and low-cost media, which allows us to see the changes in the environment because of the programming techniques that one learns in regular courses. To start in this direction, this book introduces the basic concepts of an embedded system to the reader. The citation of academic projects on robotics is expected to thrill the readers to perform hands-on demonstration to understand how the sub-systems and components in a project communicate among themselves and are being controlled.

This book introduces essential concepts and knowledge in Chaps. 1, 2, 3, 4, 5, 6, and 7, preparing the readers to undertake projects in Chap. 8 using an ABC (Acquire, Build, and Create) pedagogy within the field of robotics and embedded systems. The book is with the potential for active learning by the readers to develop problem-solving skills and create practical solutions for real-world challenges. The teaching learning methods using this book should follow: (1) formulating expected learning outcomes, (2) understanding the concept of the textbook materials, (3) skills training through realization of textbook materials using real-world examples, (4) designing a project theme, (5) making a project proposal, (6) executing the tasks of projects as per the proposal, and (7) presentation of the project report.

Young minds have always been attracted toward the term robotics due to science fiction or demonstrations of robots' applications in surrounding environments. The science of robots has become an integral part of the academic curriculum in technology programs for generating skilled manpower to address the large-scale robot deployment. It is evident that with the passage of time, scientific disciplines are no longer absolute. Inter-disciplinary applications have become essential leading to co-existence of human and robot under the domain of cobotics. The book is suitable for the readers from undergraduate or graduate programs with an interest in the intriguing area of embedded systems and robotics. It is also resourceful for self-study/out-of-school learning. The essential components required for realizing the concepts of this textbook into the physical world as presented in presented in Chap. 8 is of potential to create practical solutions to real-world problems.

With this envision, we are indebted to a number of individuals who, directly or indirectly, have assisted in the preparation of this book. In particular, we wish to extend our gratefulness to Professor Subir Kumar Saha, Department of Mechanical Engineering, Indian Institute of Technology, Delhi. Our students over the past few years have influenced not merely our thinking but also have contributed to this book. Enthusiastic students have worked with us in different projects during their graduate and postgraduate programs. Significant assistance has been provided by Tanaya Das, Dhruba Jyoti Sut, Lakhyajit Gohain, Prem Prakash Vedi, Tulika Bhuyan, Trishna Barnam, Amlan Jyoti Kalita, Debajit Chakraborty, Sandeep Choudhary, and Maibam Pooya Chanu. We express our appreciation for their support.

Tezpur, Assam, India	Nayan M. Kakoty
Ellicott City, NJ, USA	Ramana Vinjamuri
Tezpur, Assam, India	Rupam Goswami

Contents

Chapter 1
Introduction to Embedded Systems

This chapter aims to familiarize the reader with basics of embedded systems and its state-of-art applications. The fundamental components and applications of embedded systems are diagrammatically represented in this chapter. An effort is made to create a visual space in the readers' mind pertaining to the promising future orientations of embedded systems. After reading this chapter, the reader is expected to acquire the following knowledge:

1. Definition of an embedded system
2. Classification of embedded system
3. Basic components of an embedded system
4. Applications and examples of embedded system

1.1 What Is an Embedded System?

An embedded system, introduced in 1970, is a combination of software and hardware, i.e., a programmable electronic or electro-mechanical system designed for performing a specific task. The operational sequences in the system are rooted in an integrated computational system and thereby it is named *embedded*. The hardware in an embedded system consists of a programmed microcontroller or microprocessor containing memory modules, input-output interfaces, display systems, communications modules, and electronic and mechanical components. The software present in it comprises of a set of instructions for communicating and controlling the above and other sub-systems interfaced to it. It is written in a high-level programming language, compiled into machine code, and uploaded into embedded hardware, which on execution sends commands to interfacing peripheral components.

An embedded system is often confused with a general-purpose system. A general-purpose system performs multiple tasks at a time, e.g., personal computers and laptops can perform many tasks at a time such as watching a video and writing in

© The Author(s), under exclusive license to Springer Nature Switzerland AG 2024
N. M. Kakoty et al., *Introduction to Embedded Systems and Robotics*,
https://doi.org/10.1007/978-3-031-73098-6_1

word processors in addition to other programs. But an embedded system is destined to do a particular task at a time. A common and relevant example is the digital camera that captures photos, stores them in a memory card, and displays the images through the screen. Other common examples of embedded systems are washing machines, microwave ovens, MP3 players, and printers.

The domains of applications of the embedded system are industrial machines, consumer electronics, agricultural and process industry devices, automobiles, medical equipment, household appliances, airplanes, vending machines and toys, and mobile devices.

1.2   Classification of Embedded Systems

Advances in microelectronics and innovations in semiconductor technology have inspired the evolution of embedded systems. With the increase in functional requirements and better performance of embedded systems for creative technologies, the domain of embedded systems has been continuously developing since 1960.

1.2.1   Classification of Embedded System with Evolving Time

1.2.1.1. First generation: Embedded systems comprising of 8-bit microprocessors or 4-bit microcontrollers belong to the first generation. They have hardware circuits and software that includes 8085 microprocessor, electrical motor control units, and programming in machine language.
1.2.1.2. Second generation: Embedded systems comprising of 16-bit microprocessors or 8- to 16-bit microcontrollers like SCADA systems belong to the second generation.
1.2.1.3. Third generation: Embedded systems comprising of 32-bit processors or 16-bit microcontrollers are third-generation systems, e.g., Digital Signal Processors, ASICs, Intel, Pentium, etc.
1.2.1.4. Fourth generation: Embedded systems comprising of 64-bit processors or 32-bit microcontrollers belong to the fourth generation. These are powerful in terms of faster computation and higher memory, and they are built on the concept of *System on Chips* and multi-core processors, e.g., smartphone devices and mobile internet devices.

1.2.2 Classification of Embedded System Based on Performance and Functional Requirements

1.2.2.1. Stand-alone embedded systems: This category of embedded systems works alone and does not need a host system, e.g., digital cameras, microwave ovens, and video game consoles. Systems like automobile engine control units are non-stand-alone embedded systems; while they operate autonomously in controlling the engine, they also communicate with other systems such as transmission control units (TCU) and anti-lock braking systems (ABS).

1.2.2.2. Real-time embedded systems: This category of embedded systems completes a task in a particular time as instructed by the system, e.g., flight control systems, set-top boxes, and missile guidance systems. However, embedded systems in MP3 players, digital cameras, microwave ovens, washing machines, and refrigerators are not real-time embedded system.

1.2.2.3. Networked embedded systems: This category of embedded systems is connected to a network to avail the resources. Local area network (LAN), wide area network (WAN), and internet are the connected networks, e.g., home security system in LAN embedded system. Most embedded systems like digital cameras, microwave ovens, and washing machines do not require to be connected to a network.

1.2.2.4. Mobile embedded systems: This category of embedded systems is used in mobile embedded devices, and sometimes merges with stand-alone embedded systems, e.g., smartphone devices, digital cameras, and MP3 players.

1.3 Basic Components of an Embedded System

There are three main components of an embedded system: hardware, software, and real-time operating system (RTOS). Three specific categories of functions of these three components are (a) reading the input or command from the outside world; (b) processing the information; and (c) generating necessary signal as output for bringing changes in the environment. Figure 1.1 shows the basic components of an embedded system.

1.3.1 Hardware of an Embedded System

The hardware of an embedded system consists of a central processing unit (CPU), memory, and a set of input/output ports.

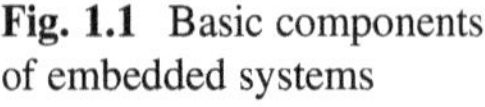

Fig. 1.1 Basic components
of embedded systems

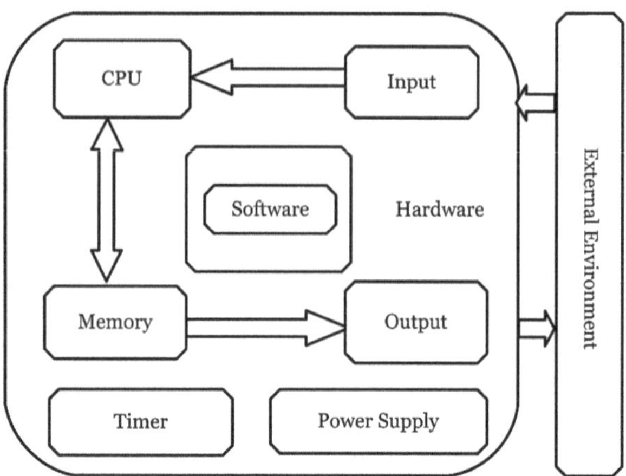

1.3.1.1 Central Processing Unit (CPU)

The CPU is responsible for processing the system inputs and taking decisions which
guide the system operation by executing the software instructions. It is the main
control unit of the system. The CPU in most embedded systems is either a microprocessor or a microcontroller, but it can also be a digital signal processor (DSP),
complex instruction set computer (CISC) processor, reduced instruction set computer (RISC) processor, or an advanced RISC machine (ARM) processor depending
on the application of the system.

1.3.1.2 Memory

The memory component is responsible for storing program code and data necessary
for system operation. The section of memory which stores information permanently
is called non-volatile memory. Read-only memory (ROM) is an example of
non-volatile memory that stores data in it even after the electrical power to the
system is switched off. Depending on the fabrication, erasing, and programming
techniques, non-volatile memories are divided into programmable ROM (PROM),
FLASH, erasable PROM (EPROM), or electrically EPROM (EEPROM) types.

The section of memory that stores information temporarily and loses its contents
when the electrical power is switched off is called volatile memory. Random access
memory (RAM) is a volatile memory. It is the main working memory of the
controller/processor where the information can be directly accessed from a memory
location. RAM is further categorized as Static RAM (SRAM) and Dynamic RAM
(DRAM) based on the technology of storing the data. The data is stored as voltage in
SRAM and as charge in DRAM. DRAM is slower and inexpensive than SRAM as it
needs to be dynamically refreshed all the time.

Solid state memory drives or commonly called SSDs are a type of computer storage device that uses flash memory to store the data electronically in non-volatile memory chips. The faster computation in SSDs makes them efficient for expanding data handling by embedded systems.

1.3.1.3   Input-Output Ports

For communication of information in between the embedded system and the external world, two types of input-output ports, communication ports and user interface ports are used.

The ports that are used for serial and/or parallel exchange of information with other devices or systems are categorized as communication ports. USB ports, printer ports, wireless RF, and infrared ports are examples of input-output communication ports. The functionality of these ports is defined specifically with respect to embedded systems.

The input-output ports that are used for exchange of information with the user are called user interface ports. Input-output ports connected to the keyboards, switches, buzzers and audio, lights, numeric, alphanumeric, and graphic displays are under this category.

Apart from these, some other hardware components included in an embedded system are power supply, timers, and counters. Power supply in an embedded system is the key component to provide stable electrical power to the system circuits. Timers and counters are used in certain applications which require delay in their functioning. Further, the amount of delay depends on the system frequency and crystal oscillator at which the embedded system is set to function.

1.3.2   Software of an Embedded System

The software of an embedded system, also called embedded software, typically represents a specialized program, i.e., a set of instructions developed for an embedded system to perform and control a specific task. The software present inside an embedded system is reprogrammable. The software is developed according to the hardware to be controlled by the specific system or application. This is in contrast to a firmware that serves as the main operating software of a computing system, but cannot be reprogrammed. On top of the firmware, the embedded software executes to provide functionality to the system.

The various tools that are generally used by a developer to develop an embedded software systems include an editor, compiler, assembler, debugger, and simulator. These tools are present in an integrated development environment (IDE). The software is written either in a high-level language like embedded C, embedded C++, JavaScript, Python or in an assembly language specific to the target controller or processor.

1.4 Characteristics of Embedded Systems

Flexibility with reference to task scheduling through multi-core embedded systems in 2010 allowed real-time systems in industrial, domestic, and scientific applications. A schematic of embedded system characteristics is shown in Fig. 1.2. While these are typical characteristics of embedded systems, a successful design must carefully consider and align them with the specific needs of the application. For such a successful embedded system design, the following characteristics are of importance.

- *Specialized operation*: The operations performed by an embedded system are specific in nature, e.g., a blood glucose meter kit is used to measure the concentration of glucose in the blood of a person.
- *Real time*: Embedded systems are able to take or read inputs continuously over time connected through peripherals, and process them to produce the corresponding desired outputs simultaneously, e.g., embedded system in a

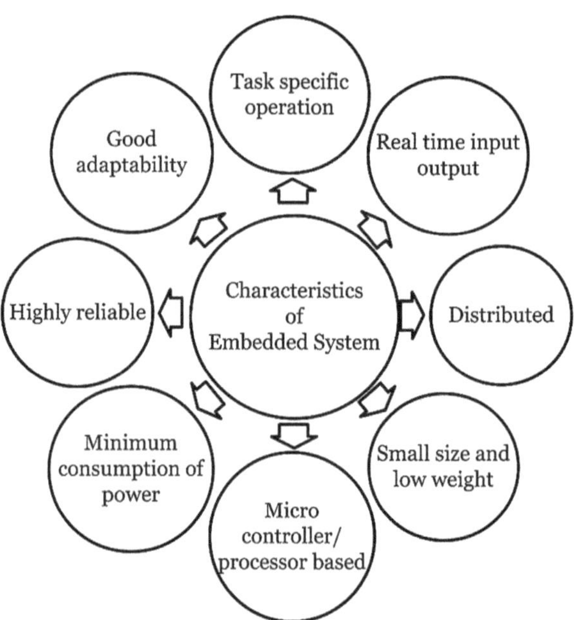

Fig. 1.2 Characteristics of embedded systems

pacemaker helps to maintain the normalcy of heart rhythms by generating electrical impulses.

- *Distributed*: Embedded systems are often part of a larger system such that many independent embedded system units together form a single embedded control unit and work together to perform a specific task, e.g., an ATM consists of a card reader chip for identifying the customer, a pin pad and a touchscreen for user interface, a cash dispensing unit, and a printer for printing the receipt. These various units are distributed embedded systems that work together to perform a particular operation.
- *Size and Weight*: Embedded systems are usually small in size and light in weight. The size and weight are important deciding factors in design of an embedded system for making them user friendly, e.g., smartphones are generally small in size and light in weight making them easier to carry.
- *Microcontroller or microprocessor-based systems*: Embedded systems are always based on a microcontroller or a microprocessor, e.g., a washing machine contains a microprocessor that controls all the operations of the machine such as checking the water level, temperature of water, washing, and setting timer.
- *Power management*: Embedded systems are designed to operate with minimum power consumption based on demands to increase the life of the system, e.g., a pacemaker system can last for an average of 6 to 7 years without recharging or replacing their batteries.
- *Adaptability:* An embedded system should be able to operate in uncontrolled environments such as a wide region with high range of temperature and pressure, e.g., military drones which are used for search and rescue operations during natural disasters (floods and earthquakes) need to operate in environments with varied wind pressure.
- *Reliability:* An embedded system should possess high reliability irrespective of its type of applications, e.g., an automatic drug delivery system has high reliability as it is used to administer drugs into patients' body.

1.5   Applications of Embedded Systems

Embedded systems play a significant role in our daily living activities starting from home to workplaces, playground to healthcare, entertainment to e-commerce, and travel to security systems. A schematic of embedded system applications is presented in Fig. 1.3 and a few examples are cited below:

Embedded systems in household: Applications of embedded systems in household appliances include microwave ovens, air conditioners, washing machines, refrigerators, dish-washers, and home automation.

Embedded systems in workplace: Applications of embedded systems in the workplace include routers, firewalls, switches, network cards, smart card readers, printers, and fax machines.

Fig. 1.3 Applications of embedded systems

Embedded systems in sports: Applications of embedded systems in the sports include smart shoes, fitness shirts, and smart watches.

Embedded systems in healthcare: Applications of embedded systems in healthcare are numerous and some of them are ECG machines, pacemaker, digital hearing aid, digital blood pressure measuring kit, tomography scanners, and prosthetic devices.

Embedded systems in entertainment: Applications of embedded systems in the entertainment industry include MP3 players, high definition television sets, video game consoles, and digital cameras.

Embedded systems in e-commerce: Applications of embedded systems in e-commerce include ATMs, cash deposit machines, and barcode readers.

Embedded systems in travel: Applications of embedded systems in travel cover cruise control system, engine control, automatic navigation systems, etc., in the automotive industry, and control systems, engine controllers, and rider in flight recreation systems in the aerospace industry.

Embedded systems in security: Applications of embedded systems in security comprise of security systems, surveillance cameras, and drones.

1.6   Future of Embedded Systems

With the advancement in the area of microelectronics, communications, sensors, memory systems, and information processing techniques, the world has become more dependent on automated systems. For better quality of life, embedded systems trained with personal profiles is of the need. With these aspects, the future of embedded systems in conjunction with robotics can be envisaged to be more user-friendly and significant. From the societal viewpoint, embedded systems will be finding more applications focused on improving the quality of life at low cost through the automation of processes. The following sections brief some of the specific technologies that will dominantly find use of embedded systems in the near future.

1.6.1 Embedded System–Based IoT

Internet of Things (IoT) is based on the concepts of algorithm (A), big data (B), and computational power (C), which, in short, may be called the ABC concept. It attempts to connect everything and everybody at any time present anywhere over the internet. With the ability of sensor technology to sense each and every activity of day-to-day life, the sensor networks are leading to generation of big data (B). Advances in algorithms (A) used for understanding the data and increase in computational power (C) with the advancement in microelectronics have enabled the decision-making process on chip in embedded systems.

1.6.2 Cyber Physical Embedded Systems

The embedded systems are supported by powerful computation and fast communication aimed at integrating the physical and the cyber worlds. Cyber physical embedded systems in robotics can be used for precision-based tasks such as in the implementation of robotic arms for medical surgery, exploration of extra-terrestrial domains, and security in biological war. Design of cyber physical embedded systems is challenging considering the issues related to privacy and flexibility due to their high level of complexity.

1.6.3 Context Awareness Embedded Systems

The meaning of context is related to circumstances. It refers to the environmental and social surroundings of a human being or an object. Such environments are dynamic in nature and require larger resources for understanding and decision-making based on context. Context-aware embedded systems utilize contextual information around them via a network of sensorized systems. This enables application of embedded systems in various devices like context-aware power-management embedded devices and context-aware home-automation devices like Alexa. This enables users to coordinate their work in order to accomplish tasks in less time and enhance efficiency. The future of context awareness embedded systems holds promises for applications in a larger market.

1.7   Advances in Embedded Systems vis-à-vis Robotics

Advances in embedded systems are creating a new era in robotic technologies. Implementing new technologies of embedded systems into robotics has increased their flexibility with being capable of performing a variety of tasks and applications with intelligence. Today's robots are more precise and consistent. The robots which were traditionally used only for pick-and-place operations in industry are now controlled by users from anywhere because of the introduction of the embedded system–based IoT.

With cyber physical embedded systems, medical diagnosis data can be analyzed remotely. This will enable robotic surgical systems with the operator being located at a distant place. It is a part of the ongoing research in current robotic surgical systems like da Vinci Robotic Surgery system where the operator needs to be present on the site of surgery.

Embedded systems equipped with context awareness capability led to the advanced humanoid systems, e.g., the first humanoid robot to receive the citizenship of Saudi Arabia can read facial expressions of human and respond accordingly. More such robotic applications could be seen in restaurants, ticket counters, healthcare systems, and entertainment industry. Further, advances in this area will enable friendly robot door-to-door service applications.

Chapter 2
Introduction to Robotics

This chapter will introduce the reader to preliminary understanding of "What is a robot and robotics?" starting with a brief history of robotics since its inception. After going through this chapter, the reader is anticipated to acquire the following knowledge:

1. Brief history of robotics
2. The laws of robotics
3. Basic terminologies and fundamental concepts of robotics
4. Applications of robotics

2.1 History of Robotics

The term "robot" first appeared in the play "Rossum's Universal Robot (R.U.R.)" by the Czech writer, Karel Capek, and has been in use since 1921. It originated from the Czech word "robota" which means "forced labor." The first robot was an industrial robot designed and patented by J. Engelberger and George C. Devol in 1954. They started Unimation Robotics Company during 1958 and manufactured the commercial version known as *Unimate*. This robot was first used in the automobile company General Motors for automation of die-casting and spot-welding operations. Figure 2.1 shows a brief schematic representation of the history of robotics. Inclusion criteria for the events of the history of robotics brief history are as follows:

1. Coining of Key Terminology: Events where key terms and concepts in robotics were first introduced.
2. Technological Firsts and Innovations: The first occurrences of significant technological advancements in robotics.
3. Pioneering Autonomous Systems: Key developments in autonomous robotic systems.

N. M. Kakoty et al., *Introduction to Embedded Systems and Robotics*,
https://doi.org/10.1007/978-3-031-73098-6_2

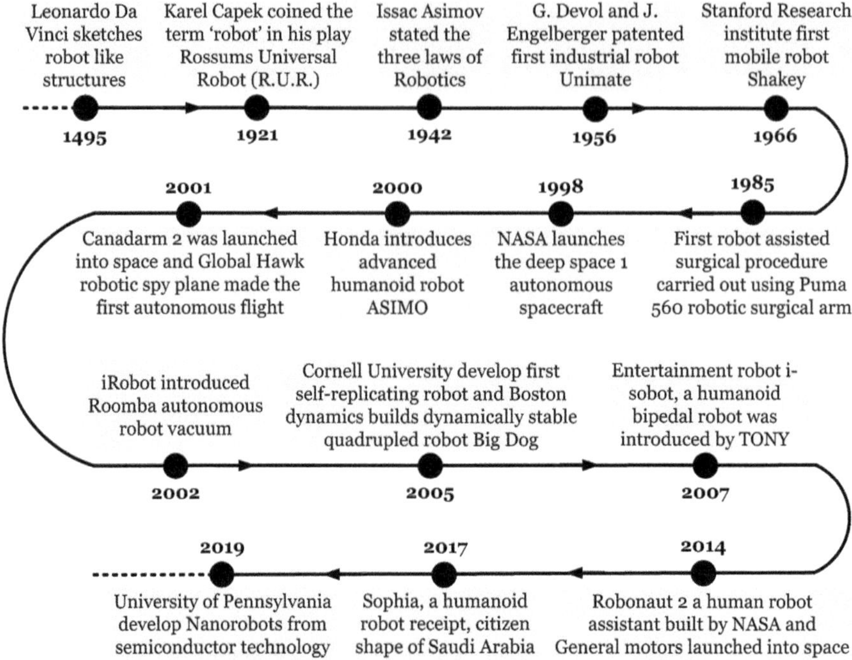

Fig. 2.1   Timeline diagram of history of robotics

2.2   Laws of Robotics

Asimov's three laws of robotics, which were shaped in the literary work of Isaac Asimov (1920–1992), define a crucial code of behavior that fictional autonomous robots must obey as a condition for their integration into human society. They are known as the "Three Laws of Robotics":

1. A robot may not injure a human being, or, through inaction, allow a human being to come to harm.
2. A robot must obey the orders given to it by human beings except where such orders would conflict with the First Law.
3. A robot must protect its own existence as long as such protection does not conflict with the First or Second Law.

Although efforts have been made to follow the laws of robotics, there are no involuntary methods of implementing them.

2.3   Terminologies and Basic Concepts

Robot: Robot Institute of America in 1979 defined "robot" as a reprogrammable, multi-functional manipulator designed to move material, parts, tools, or specialized devices through variable programmed motions for performing a variety of tasks.

In the current scenario, intelligence that allows a machine to learn and react in unknown environment to accomplish variety of tasks is one of the most eagerly searched characteristics in robots. The basic components of a robot are as in Fig. 2.2.

Robotics: It is the science of robots. Humans working in this area are called roboticists. Robotics is highly interdisciplinary in nature. The major disciplines involved in robotics are presented in Fig. 2.3.

Position: The translational location of an object is known as position. An object's position is defined by three translational locations in space. The position of a point P in a Cartesian coordinate system {X, Y, Z} is shown by the vector P in Fig. 2.4.

Orientation: The rotational location of an object is known as orientation. Figure 2.5 shows the orientation of an object using the angles of the objects with the axes of a coordinate system.

Pose: Position and orientation taken together is known as pose.

Degrees of freedom (DoF): The number of axes along which and about which an object can translate and rotate freely is defined as the degrees of freedom of the object.

In free space, a point object has three degrees of freedom (three translational movements) while a solid object has six degrees of freedom (three translational and three rotational movements).

Link: A rigid piece of material connecting joints in a robot is known as link.

Joint: The device which allows relative motion of one link with reference to another link is known as a joint. Figure 2.6 shows the links and joints present in a robotic arm. Depending on the types of application, there are different types of joints used in robotics as shown in Fig. 2.7.

End-effector: The tool, gripper, or other device mounted at the end of a manipulator for accomplishing useful tasks is known as end-effector.

Fig. 2.2 Basic components of a robot

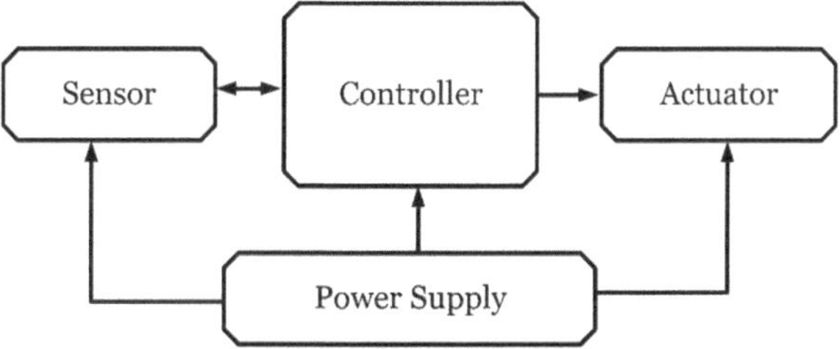

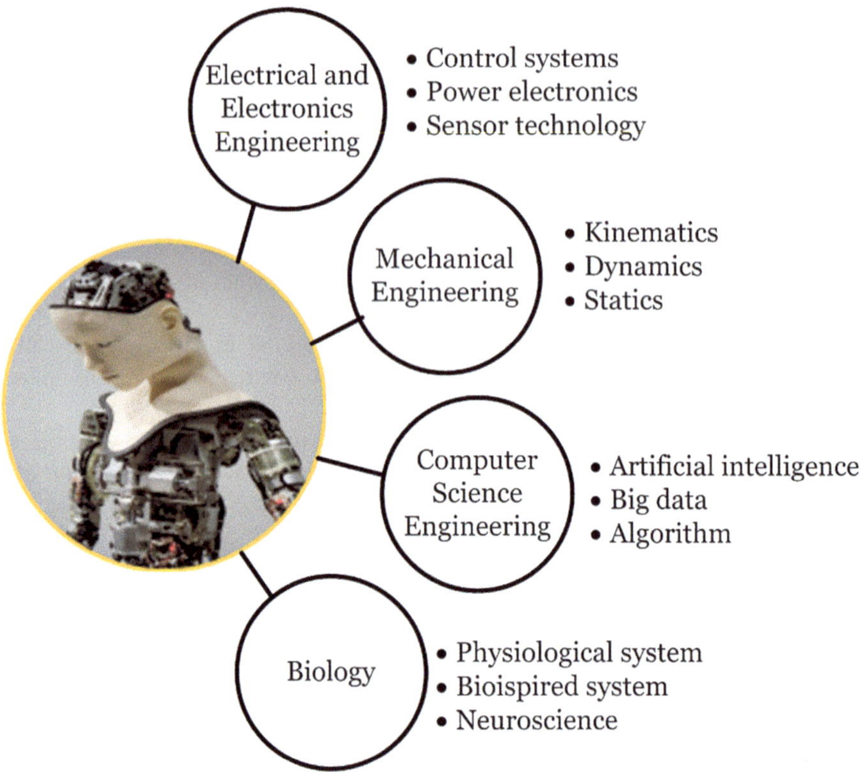

Fig. 2.3 Different disciplines involved in robotics

Fig. 2.4 Position of point P in Cartesian coordinate {X, Y, Z}

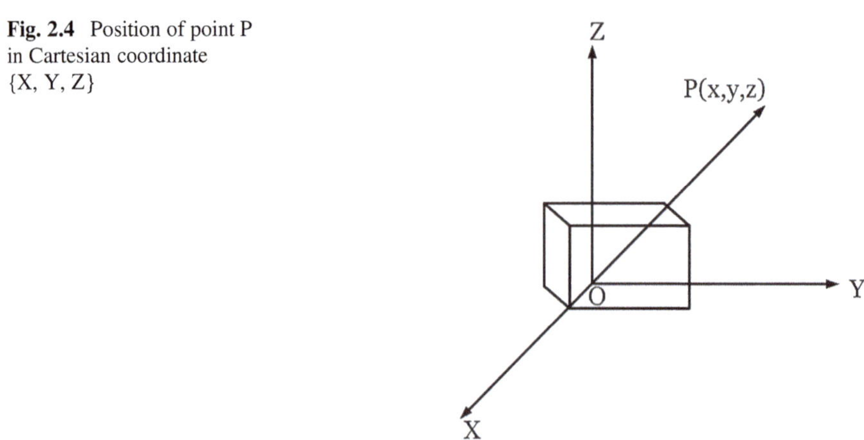

Workspace: The volume in space that a robot's end-effector can reach, both in position and in orientation, is known as workspace. In Fig. 2.8, the volume enclosed by the lines surrounding the robotic arm shows its workspace.

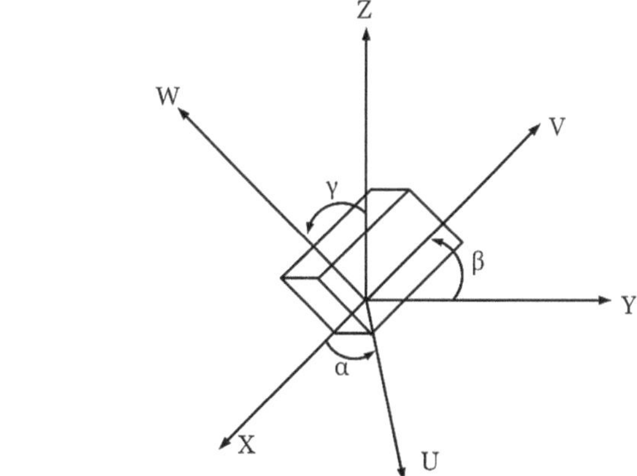

Fig. 2.5 Orientation of an object expressed using angles α, β, and γ with the coordinate axes

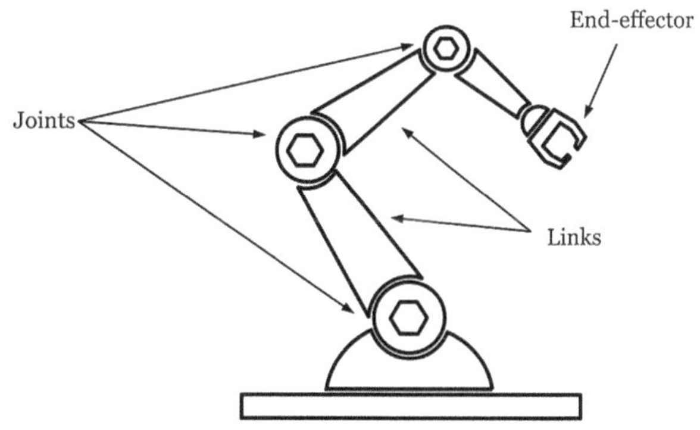

Fig. 2.6 Links, joints, and end-effector of a robotic arm

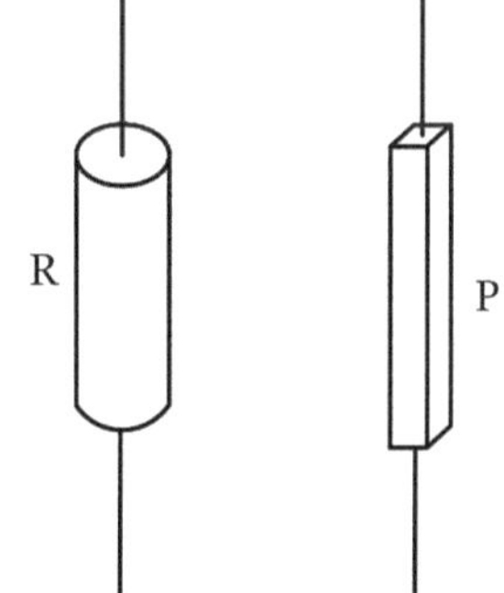

Fig. 2.7 Different types of joints: (**a**) revolute joint and (**b**) prismatic joint

Kinematics: The discipline of engineering concerned with the study of motion of objects without considering the forces which cause the motion is known as kinematics.

Dynamics: The discipline of engineering concerned with the study of forces and torques and their effect on motion is known as dynamics.

Fig. 2.8 Schematic of workspace of a robotic arm

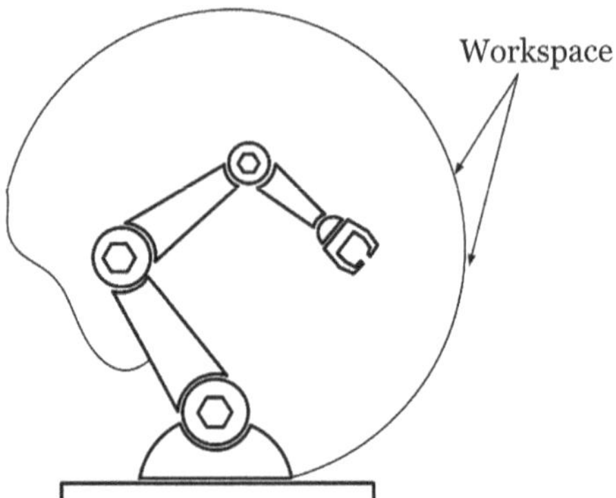

Actuator: Actuator of a robotic system is the component that brings the changes in the pose of other components in the system. The typical energy sources for these changes are electric, hydraulic, fluid pressure, pneumatic pressure, etc.

Sensor: Sensor is a device or a module that converts the changes of a physical parameter or phenomenon into an equivalent electrical quantity.

Co-robot: Co-robot or collaborative robots are robots that interact with humans with close proximity to each other or in a commonly shared work space.

2.4 Applications of Robots

Since inception, robots have been in practice for their inherent advantages in terms of accuracy, safety, precision, and robustness. Robots can be categorized under the following categories depending on their type of application:

2.4.1. Industrial robot: Industrial robots are the ones deployed for industrialized manufacturing and shipment of goods. Industrial robots can be either in manufacturing or in logistic applications.

> *Manufacturing robots*: These robots perform programmed repetitive tasks in a customized and well-defined environment such as in an assembly line in a factory. The commonly used manufacturing robots are articulated arms or robotic arms that are specifically created for operations like material handling, painting, welding, assembling, disassembling, picking, and placing. KUKA robot is one of the most popularly used manufacturing industrial robots.

> *Logistic robots*: Robots finding applications for storage and transportation of goods in industries are logistics robots. Because of the inherent advantages of these robots and increasing need of rapid parcel shipments in e-commerce industries, use of logistic robots is increasing every day. These robots are mostly mobile and automated vehicles that operate in warehouses and storage facilities to transport goods.

2.4.2. Service robots: Robots that provide assistance to humans in their day-to-day activities including domestic, office, space exploration, and hospitals are under the category of service robots.

Domestic or household robots: These service robots are employed at home for doing household tasks with ease, e.g., assistive robots for elderly people and floor-cleaning robots.

Medical robots: Medical robots are service robots employed in hospitals and medical facilities. Medical robots need to be extremely accurate and precise in their operations, e.g., surgical robots like the da Vinci surgical system and rehabilitation robots like prosthetic limbs.

Military robots: Robots finding applications in military service fall under this category. This category of robots needs to be very robust in their operations irrespective of their work space, e.g., autonomous or remote-controlled bomb discarding robots, military drones, and underwater vehicles.

Entertainment robots: Entertainment robots are the type of service robots used for enrichment of intelligent quotient through fun, e.g., Aibo—a robotic dog, humanoid robots like QRIO and RoboSapien.

Space robots: These types of robots are deployed for extra-terrestrial space exploration and to assist astronauts in space stations, e.g., Mars Rover Curiosity and humanoid robot like Robonaut.

Educational robots: Robots used for realization of text-book concepts through visualization of changes in the physical world belong to this category, e.g., Lego Mindstorms robotic systems, Vex robotic system design, and Thymio. Software tools like RoboAnalyzer (developed in the Indian Institute of Technology, Delhi) and GraspIt! (developed in the Columbia University) are also used for educational purposes. There also exist educational robots that are used to carry out teaching functions, e.g., ABii robot (developed by VAN Robotics), NAO robot (developed by SoftBank Robotics), and EMYS (developed by Flash Robotics).

Chapter 3
Sensors, Actuators, and Circuits in Robotics

There are a number of sensors like proximity, infrared, temperature, and pressure that one encounter often during daily living. Human body is equipped with five different types of sensors: eyes that detect light energy, ears that detect acoustic energy, tongue and nose that detect certain chemicals, and skin that detects pressure and temperatures. The eyes, ears, tongue, nose, and skin receive these variables from the environment and transmit signals to the brain which controls the response. For example, when you touch a hot plate, it is your brain that informs you it is hot and, therefore, you take your hands off it. In this case, the skin works as the sensor, brain as the controller, and hand as the actuator.

This chapter describes the basic concepts of sensors, actuators, and controller circuits used in robotics. On completion of this chapter, readers are expected to have gained knowledge on the following:

1. Sensors and actuators used in robotics
2. Skills to implement circuits used in robotics

3.1 Basic Concepts of Sensors

A sensor is a device used to measure events or changes in its environment and convert them into an equivalent electrical signal. Robots use different types of sensors to acquire information. Based on the sensors' inputs about the workspace and surrounding environment, a robot takes decisions on how to act.

The input to a sensor is a physical quantity like pressure, temperature, humidity, light, sound, and touch and its outputs are in the form of electrical signal. The output signals can either be analog or digital, depending on the type of sensor. These signals are fed to a controller in a robot wherein they are processed to understand the robot's environment.

N. M. Kakoty et al., *Introduction to Embedded Systems and Robotics*,
https://doi.org/10.1007/978-3-031-73098-6_3

3.2   Proprioceptive/Exteroceptive Sensors

Proprioceptive sensors measure values that are internal to the system (or robot); for example, an encoder reads the speed of a motor, a potentiometric sensor senses the joint angles in a robotic arm, and a current sensor senses the electrical charge status in a battery.

Exteroceptive sensors acquire information from the robot's environment; for example, infrared or ultrasonic sensors sense the presence of obstacles in the robot's path, gas sensors sense the chemical composition of gases in the robot's environment, and pressure sensors measure the pressure experienced by a prosthetic hand.

3.3   Sensors Used in Robotics

Sensors commonly used in robotics are based on the principle of changes in electrical parameters like resistance, capacitance, and inductance as a function of the changes in the environmental physical quantities. The physical quantity to be measured by a sensor is known as a measurand. The sensors respond to the measurands and produce proportional outputs. Sensors that use changes in resistance, capacitance, and inductance due to the changes in the measurand are known as resistive, capacitive, and inductive sensors, respectively. Most of these sensors are manufactured using semiconductor fabrication technology. Some of the advanced and miniaturized sensors are manufactured using micro-machining technology.

3.3.1   Infrared Sensor

An infrared sensor senses the surrounding environment by detecting infrared radiation. Infrared sensors are of two types.

3.3.1.1   Active Infrared Sensors

Active infrared sensors or proximity sensors can emit as well as detect infrared radiation. When an object is in proximity of the sensor, a beam of infrared light emitted from the sensor's LED is reflected back from the object and gets detected by the receiver. It is commonly used in robots for obstacle detection.

3.3.1.2 Passive Infrared Sensor

Passive infrared (PIR) sensor can detect infrared radiation but cannot emit it. When an object in motion that emits infrared radiation comes under the sensing range of the sensor, two pyroelectric elements present inside the sensor measure the difference in the amount of infrared radiation levels between them leading to a change in the output voltage triggering the detection. PIR sensors are mostly used for detecting presence of humans. The basic element of a PIR sensor is a P-N junction which works based on the principle of recombination of electron-hole pairs (EHPs). Infrared light radiation from an object onto the P-N junction impacts the recombination near the depletion region as presented in Fig. 3.1.

3.3.2 Ultrasonic Sensors

An ultrasonic sensor is used for measuring the distance or detecting a target object in the surrounding environment. It comprises of a transmitter, which transmits ultrasonic sound waves to the surrounding, and a receiver, which receives the reflected sound wave from an object. Ultrasonic sensors are independent of ambient light.

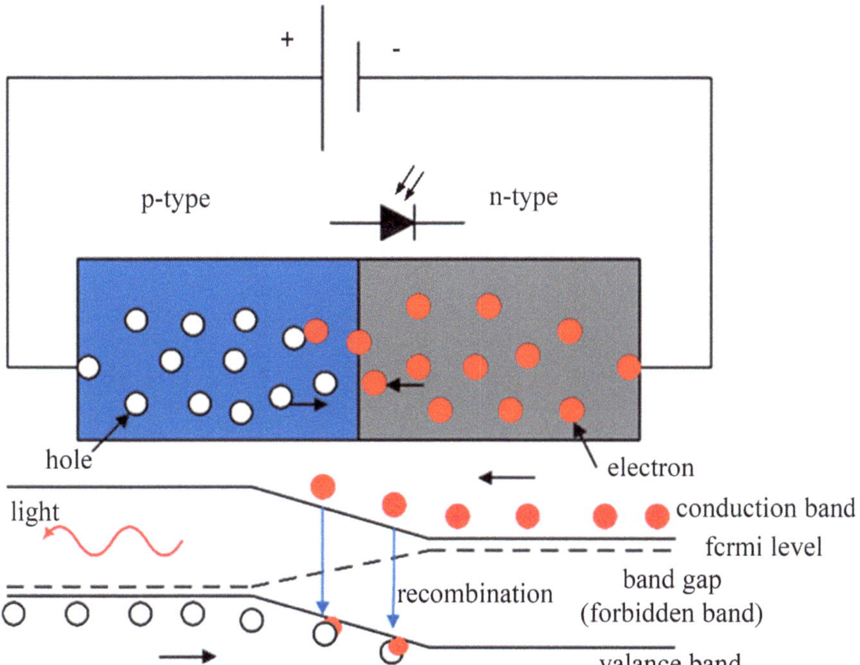

Fig. 3.1 Recombination of carriers in the basic component of a PIR sensor

3.3.3 Light Sensor

A light sensor detects light and converts it into an electrical signal. The light energy may be in the part of either visible or infrared light spectrum. The light sensor generates a voltage difference corresponding to the light intensity. The two commonly used light sensors in robotics are light-dependent resistors and photovoltaic cells.

3.3.4 Color Sensor

A color sensor detects the color of an object and converts into a frequency which is proportional to the intensity of the color. It is used for making color-sorting robots to distinguish different colors.

3.3.5 Rotary Encoder

A rotary encoder is a sensing device that detects position and speed, and converts the angular motion or position of a shaft or axle into an electrical signal. The electrical signal can be either analog or digital. It is used in robotics to serve as a feedback system for position and speed control (Fig. 3.2).

Fig. 3.2 Rotary encoder

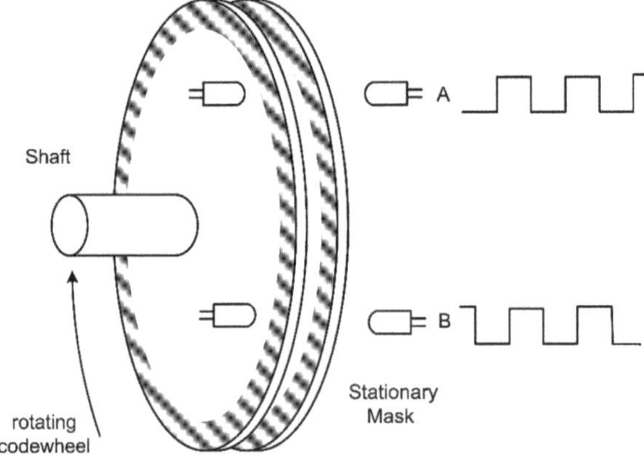

3.3.6 Accelerometer

An accelerometer, based on micro electro mechanical system, is manufactured using micro-machining technology. It is used to measure the change in acceleration and tilt angles of an object. The sensor responds to both static and dynamic acceleration (Fig. 3.3).

3.3.7 Touch Sensor

A touch sensor gets triggered on by being in physical contact with other objects. It works based on capacitive sensing principle. Touch sensors are commonly used in the development of prosthetic hands and industrial grippers.

3.3.8 Pressure Sensor

A pressure sensor measures pressure using resistive sensing. It converts the force applied on the sensing area of the sensor into an electrical energy in the form of voltage. Most pressure sensors work on the principle of piezo-resistive sensing.

3.3.9 Thermal Sensor

A thermal sensor detects any changes in the surrounding temperature, and generates a voltage difference corresponding to the temperature change. It functions based on resistive sensing principle. LM34, LM35, TMP35, TMP36, and TMP37 are a few frequently used thermal sensors.

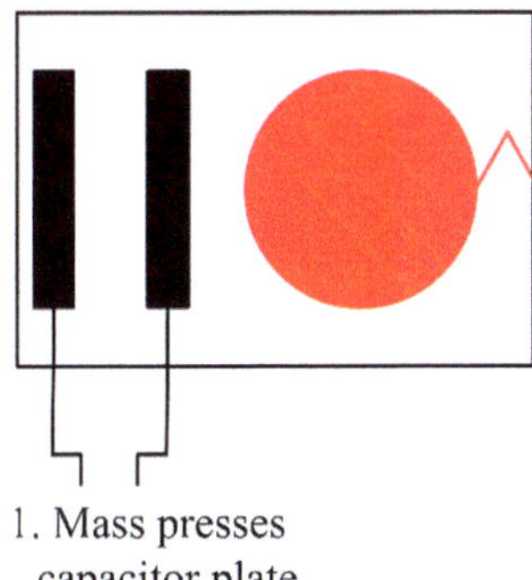

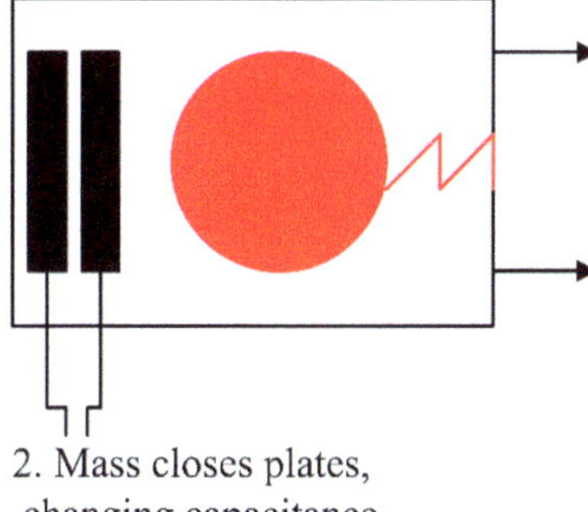

Fig. 3.3 Capacitive accelerometer

3.3.10   Position Sensor

A position sensor, as in a global positioning system, is used to determine the approximate location of a robot. It receives and processes signals from the satellites that orbit the earth. It is used in robots for positional feedback, e.g., Neo 6M TTL GPS module.

3.3.11   Magnetic Sensor

A magnetic sensor measures the magnetization of a magnetic material like a ferromagnet and gives the direction of the magnetic field at a point in space. They act similar to that of a magnetic compass. It is used for robots in navigation systems, e.g., Philips KMZ51 (Fig. 3.4).

3.3.12   Emotion Sensor

An emotion sensor interprets human facial expressions. It is used to build robots like humans, which require detection of emotions, e.g., B5T HVC face detection sensor.

Fig. 3.4 Bridge circuit for magnetic sensing

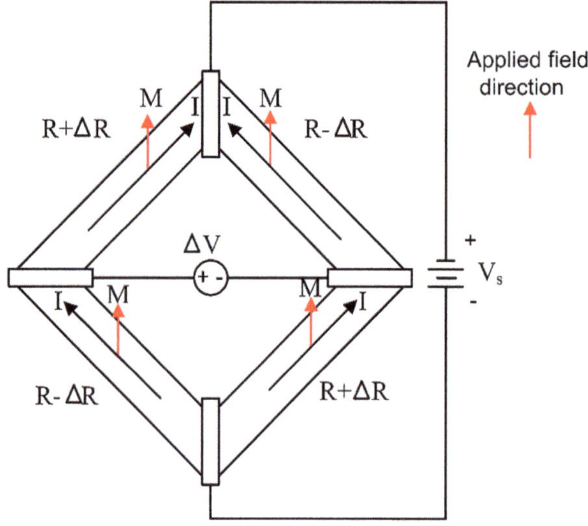

3.3.13 Sound Sensor

A sound sensor detects sound waves from the environment and converts them into an electrical energy in the form of voltage that is proportional to the sound level. The generated voltage difference by the sensor being minimal requires further amplification to produce a measurable voltage change. Sound sensors are usually used in robots to recognize speech or to develop simple clap-based robots, e.g., microphone sound sensor.

3.3.14 Water Flow and Rain Drop Sensor

A water flow sensor measures flow rate of water in a channel. It consists of a water rotor and a hall effect sensor that generates an electrical pulse on every revolution of the rotor measuring how much water has flown through it. It is used in irrigation robots, e.g., YFS201 water flow sensor.

A rain drop sensor detects waterdrops, and converts the change in resistance caused due to the raindrops fallen on the sensor into a voltage. It is used in robots employed for monitoring weather conditions, e.g., rain weather sensor module.

3.3.15 Gas Sensor

A gas sensor is used to sense different types of gases and measure their concentrations. It detects the gas and converts the corresponding resistance change inside the filament of the sensor into a voltage signal. It is used in gas detection robots for sensing toxic and flammable gases, e.g., MQ2 smoke sensor, MQ5 CO2 sensor.

3.3.16 Biosensors

Biosensors are innovative analytical devices that detect the presence or concentration of biomolecules called analytes. The input analytes react with a biological sensing membrane. On reaction with analytes, the membrane undergoes changes in its electrical parameters and thereby quantifies using a transducer. This electrical signal is further analyzed in a signal processing unit. Applications of biosensors are commonly found in medical sciences and health care. In robotics, biosensors are mostly used in assistive robotics and in man-machine interfacing. More applications of biosensors for military robots are under research.

3.4   Performance Characteristics of Sensors

Accuracy: It is defined as the difference between measured value and true value. The error in measurement is specified in terms of accuracy. It is defined in terms of % of full scale or % of reading.

Precision: It refers to the closeness of output values with which the sensor can measure a physical quantity. For e.g., a high-precision pressure sensor will give very similar readings for repeated measurements of the same pressure, say, 100.2 kPa, 100.3 kPa, and 100.2 kPa.

Resolution: It is the minimum change in input that can be sensed by the sensor. For e.g., if resolution of a pressure sensor is 0.01 kPa, it can detect pressure changes as small as 0.01 kPa.

Sensitivity: It is defined as the change in output response to change in input response of the sensor. For e.g., a sensitive pressure sensor will produce a noticeable electrical signal change for a small pressure variation, say 0.01 kPa, indicating its high sensitivity to small changes.

Linearity: Linearity is the maximum deviation between the measured values of a sensor from ideal value or relationship between input and output signal variations. The linearity of a sensor indicates how closely it adheres to ideal behavior.

Repeatability: It is defined as the ability of a sensor to produce the same output every time when the same input is applied, and all the physical and measurement conditions are kept the same including the instrument and ambient conditions. Repeatability ensures that the same sensor, under the same conditions, produces consistent and reliable readings.

Reproducibility: It is defined as the ability of sensor to produce the same output when same input is applied. Reproducibility confirms whether an entire sensor can be reproduced in its entirety.

Range: Difference between the minimum and maximum output provided by a sensor.

Response Time: It expresses the time at which the output reaches a certain percentage (for instance, 95%) of its final value in response to a step change of the input.

Saturation: It is defined as the state in which the limiting value of the sensor range becomes the output value of the sensor. This happens when the quantity to be measured by the sensor is larger than the dynamic range of the sensor.

3.5   Basic Concepts of Actuator

To make something move, we need to apply a force or a torque on it. Actuators are the generators of the forces or the torques that robots employ to move themselves and other objects. Actuators are the muscles of a robot. All actuators are energy-consuming mechanisms that convert various forms of energy into mechanical work.

The mechanical linkages and joints of the robot manipulator are driven by actuators which may be pneumatic or hydraulic or electric. These actuators may connect directly to the mechanical joints or may drive mechanical elements indirectly through gears, chains, wires, tapes, or ball screws.

3.6   Types of Actuators

There are basically two types of actuators in robotics based on the motion produced:

3.6.1   Rotational Actuators

Rotational actuators produce rotational motion of a robotic link or a joint with respect to its adjacent links or joints. The motion is generated by transforming the electrical energy into a rotating motion. There are two main mechanical parameters associated with rotational actuators: torque and rotational speed. Electrical motors like AC motors, DC motors, servo motors, and stepper motors are rotational actuators.

3.6.2   Linear Actuators

Linear actuators produce linear motion, i.e., motion along one straight line of a robotic link or a joint with respect to its adjacent links or joints. DC linear actuator, solenoids, muscle wire, pneumatic, and hydraulic cylinders are linear actuators. Linear actuators are mainly specified by three parameters: the minimum and the maximum distance that the joint or link can move, force required for the movement, and the speed of movement.

3.7   Some Actuators Used in Robotics

3.7.1   AC Motor

An AC motor is a type of rotational actuator that converts alternating current (AC) into a mechanical power. It is driven by the principle of electromagnetic induction which states that on providing alternating current to the motor's stator windings, a rotating magnetic field is developed around the stator windings. Due to this rotation, the rotor part of the motor experiences an induced current, thus resulting in a rotating magnetic field outside the stator and producing a torque

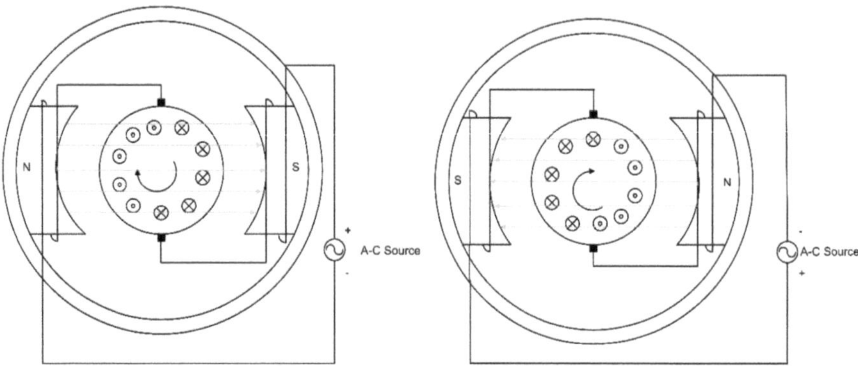

Fig. 3.5 Schematics of an AC motor

(Lorenz law) which rotates the motor. A commonly used AC motor is a three-phase induction motor. AC motors are mostly used in industrial robotics for high torque applications. Figure 3.5 represents schematics of an AC motor.

3.7.2   DC Motor

A DC motor is a type of rotational actuator that converts direct current (DC) into a mechanical power. It operates on the principle of electromagnetic induction which states that when a current-carrying conductor is placed in a magnetic field, it experiences a mechanical force. The angular motion of rotating shaft of the motor is measured using encoders or potentiometers. DC motors are commonly used actuators in robotic applications such as in wheels of a robot.

3.7.3   Servo Motor

A servo motor is a type of rotational actuator comprised of DC motor, gears, driver circuit, encoder, and a rotary potentiometer. It is a three-terminal motor with power, ground, and control lines. The driver circuit and potentiometer work in unison to activate the motor and stop the output shaft at a specified angle. The encoder stores the last angular position of the shaft and restarts from the stored angular position when electrically powered. The potential to the control line controls the shaft rotation to the new position. Servo motors are used in robotic applications which need precise positioning. A schematic of a servo motor is shown in Fig. 3.6.

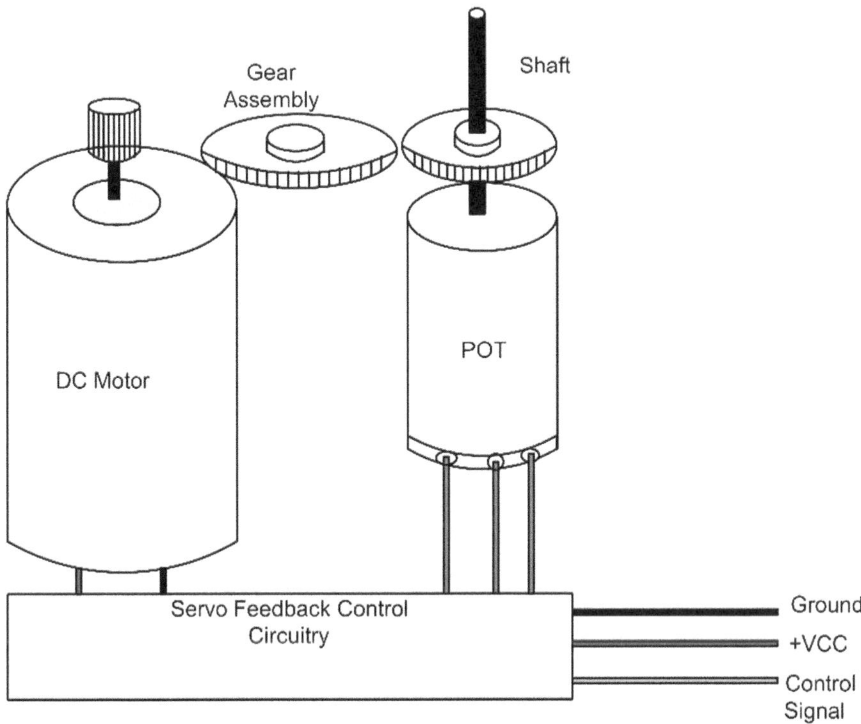

Fig. 3.6 Schematic of a servo motor

3.7.4   Stepper Motor

A stepper motor is a type of rotational actuator that rotates in small angular steps. It works on the principle of electromagnetism. A stepper motor consists of a stationary part called stator and a moving part called rotor. When current flows in the coils of the stator by energizing one or more of the stator phases, a rotating magnetic field is developed and the rotor which is a permanent magnet gets aligned with the direction of the generated field. As a result, the rotor starts rotating with the rotating magnetic field in steps by a fixed number of degrees to finally achieve the desired position. Stepper motors are used in robotic applications where discrete steps or angles of orientation are required. Figure 3.7 shows the schematic of a stepper motor.

3.7.5   DC Linear Actuator

A DC linear actuator is a type of linear actuator that produces a linear movement transforming rotation of a DC motor into a linear motion. It consists of a DC motor, a

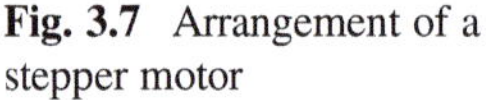

Fig. 3.7 Arrangement of a stepper motor

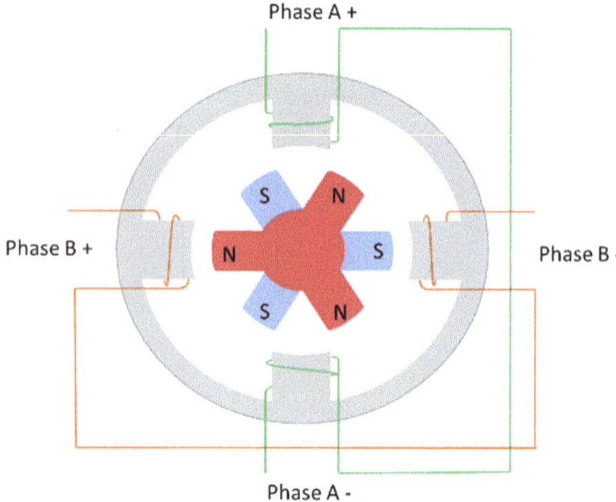

Fig. 3.8 Arrangement of a linear actuator

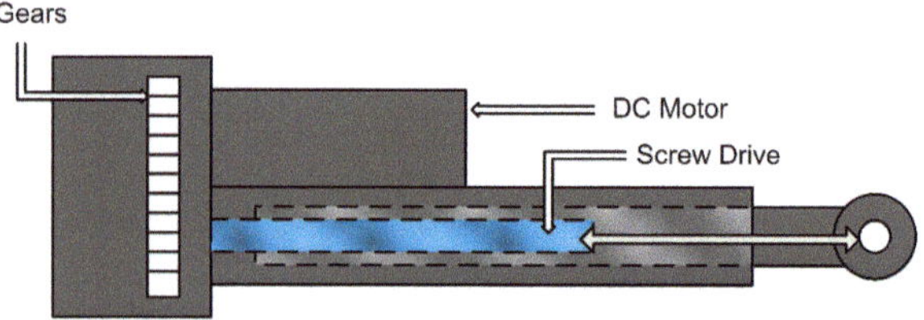

set of gears, and a lead screw. The concept behind the working of a DC linear actuator is of an inclined plane where the lead screw serves as a ramp generating a small rotational force that acts along a larger distance to move the load. DC linear actuators are added with a linear potentiometer for providing a linear position feedback. It is used in robotic applications for lifting and tilting of objects or machines. A schematic of a DC linear motor is shown in Fig. 3.8.

3.7.6 *Pneumatic Actuator*

A pneumatic actuator is a type of actuator that produces rotational motion and linear motion by using compressed air. It comprises of a piston, cylinder, and a valve or a port. When compressed air enters into the cylinder through the valve, pressure builds up inside the cylinder which results in either a controlled linear motion or a rotary motion of the piston. It is used in automation industries. A schematic of a pneumatic actuator is shown in Fig. 3.9.

Fig. 3.9 A pneumatic actuator

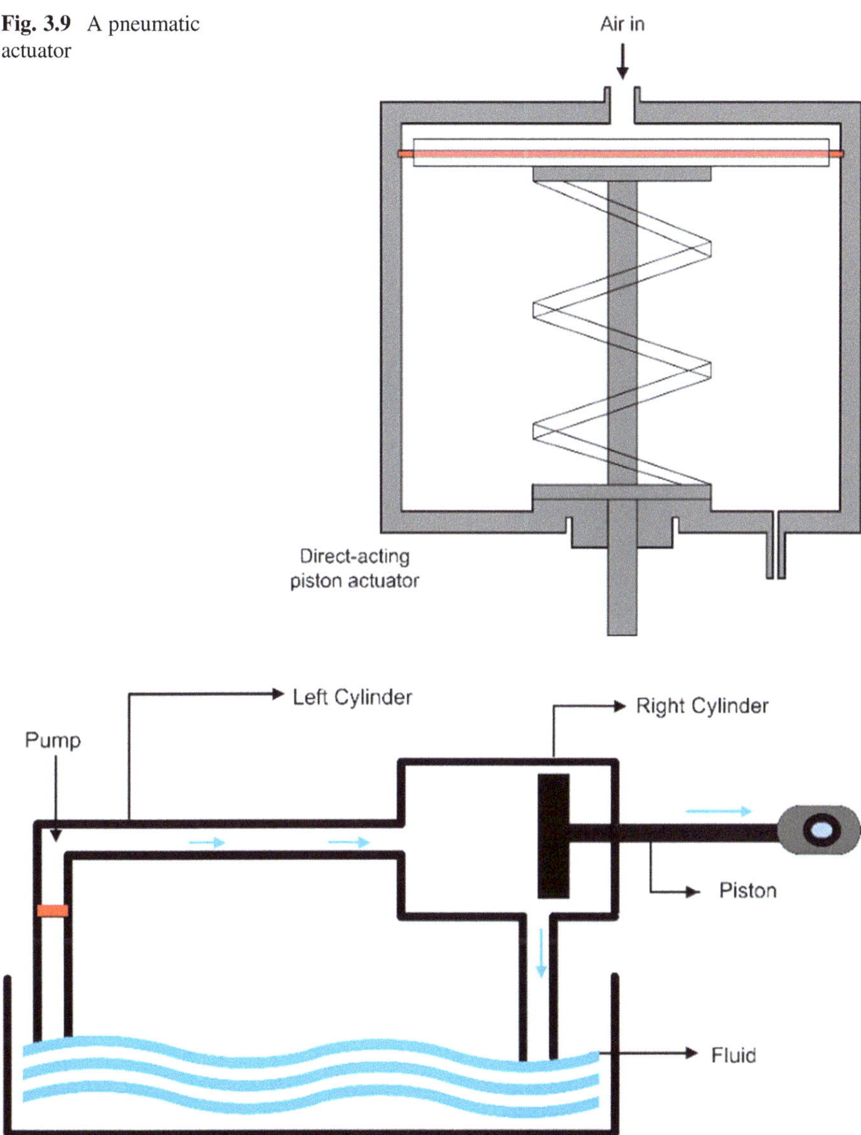

Fig. 3.10 A hydraulic actuator

3.7.7   Hydraulic Actuator

A hydraulic actuator is a hydraulic cylinder with a piston that uses a pressurized fluid, typically oil, to generate a force which eventually produces a linear displacement. It is used in robotic applications which require a force of higher magnitude. A schematic of a hydraulic actuator is shown in Fig. 3.10.

3.8   Circuits Used in Robotics

The general electronic circuits used in robotics vary depending upon the specific application of the robot. However, some of the very commonly used circuits are discussed in the following sections.

3.8.1   Power Supply Circuit

It is one of the critical components of any robotic system for its proper functioning. This type of circuits acts as the energy source for all the units in a robotic system. It is used to convert electrical power from a source to the required voltage, current, and frequency as per the specifications of the robot's components. Accurate power supply is the first criterion for any of the units in a robotic system to work properly. All power supplies have a power input connection, which receives energy in the form of electric current from a source, and one or more power output connections which deliver current to the robot's components.

The source of power may come from an electric power grid, such as electrical outlets or energy storage devices which include batteries, fuel cells, generators, and solar cells. The input and output are usually hardwired circuit connections. Some power supplies are separate standalone pieces of equipment, while others are built into the load appliances that they power. Specific circuits for efficient power management are part of power supply circuits.

3.8.2   H-Bridge Driver Circuit

H-bridge is an effective method for driving electrical actuators in controlling speed and direction. An H-bridge circuit contains four switching elements, either transistors or MOSFETs, with the actuator to be controlled at the center forming an H-like configuration. By activating two particular switches at the same time, the direction of current flow through the actuator can be changed. Also, changing the duration of switch activation and amplitude of current passage, the speed of the actuator can be controlled.

3.8.3   Amplifier Circuit

An amplifier circuit is used to increase the magnitude of an applied input signal in a system. Amplifier is the generic term used to describe a circuit which produces an amplified version of its input signal. The primary objective of an amplifier circuit is to boost the current or voltage in any circuit at any stage. It is also called buffer circuit in some applications. Buffer circuits are mainly used where a signal has quite low current input which needs to be increased for maintaining the same voltage levels. They draw current from power source and add it to the signal.

Chapter 4
Microcontrollers in Robotics

Microcontroller is a system-level component of an electronic circuit through which output based on the input received from the physical world through input-output devices like sensors is determined. It can be assumed to be analogous to the human brain. It has created great impact on mankind through its various applications—in household applications like washing machine, microwave oven, and air-conditioner; office applications like photocopy machines, computers and printing machines; industrial applications like manufacturing robots, automated warehouse trolleys, weighing machines, and automated teller machines. All the applications of an embedded system mentioned in Sect. 1.5 find the use of microcontrollers.

This chapter provides a basic understanding of the fundamentals of microcontroller and the reader is expected to understand the following after reading this chapter:

1. Basic concepts in microcontroller
2. Programming a microcontroller
3. Commonly used microcontrollers in robotics

4.1 What Is a Microcontroller?

Microcontroller is an embedded system comprising of microprocessor with on-chip RAM, ROM, timer, input-output port, and serial communication port. Microcontrollers are more suitable in a variety of electronic applications due to low cost, less space requirement, low power consumption, and high computing capacity. A schematic representation of the units in a microcontroller is presented in Fig. 4.1.

N. M. Kakoty et al., *Introduction to Embedded Systems and Robotics*,
https://doi.org/10.1007/978-3-031-73098-6_4

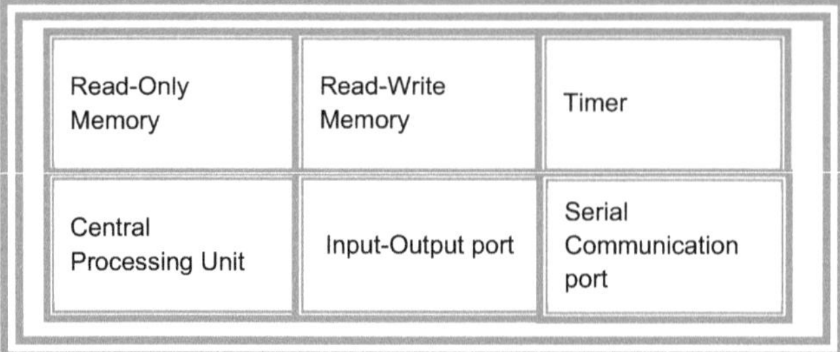

Fig. 4.1 Schematic representation of the units in a microcontroller

Unlike microprocessors, microcontrollers possess peripheral devices like analog-to-digital converters, memory registers, and timers housed in a single chip. The CPU performs input-output read/write and memory read/write operations using three sets of communication lines: data bus, address bus, and control bus. A bus is a set of communication lines.

Memory: Memory in a microcontroller stores data and instructions necessary for its operation. There are two types of memory: (a) non-volatile memory where data stored is permanent and (b) read-write memory where data stored can be rewritten. The commonly used non-volatile memory in a microcontroller are read-only memory (ROM) and a FLASH memory. For read-write memory, a random-access memory (RAM) is used.

Instruction Set: An instruction set in a microcontroller is a set of commands used to instruct the microcontroller in a machine language for carrying out specific tasks. It is specific to a microcontroller.

4.1.1 How Are Microcontrollers Classified?

Microcontrollers are classified on the basis of their number of bits, i.e., width of the data bus, memory device, instruction set, and memory architecture.

(i) *According to the size of data bus or number of bits*:

   Microcontrollers are classified as 8-bit, 16-bit, and 32-bit based on the size of data bus, i.e., number of communication lines present in it. 8-bit microcontrollers are able to manipulate 8-bit data ranging from 0×00-0xFF ($2^8 = 255$ numbers) during every clock cycle. The examples of 8-bit microcontrollers are Intel 8031/8051, PIC1x, and Motorola MC68HC11 families.

   16-bit microcontrollers perform with greater precision and performance as compared to 8-bit microcontrollers. 16-bit microcontrollers have 16-bit data width with a range of 0x0000-0xFFFF ($2^{16} = 65535$ numbers). Some examples of 16-bit microcontrollers are 8051XA, PIC2x, Intel 8096, and Motorola MC68HC12 families.

32-bit microcontrollers use 32-bit data bus and can communicate 32-bits in parallel. This results in much faster operations and higher precision compared to 8-bit or 16-bit microcontrollers. 32-bit microcontrollers are able to manipulate 32-bit data ranging from 0×00000000-$0xFFFFFFFF$ ($2^{\wedge}32 = 4294967295$) during every clock cycle. Some examples are Intel/Atmel 251 family, PIC3x.

(ii) *According to memory devices*:

Microcontrollers are divided into two types based on intrinsic and extrinsic memory as embedded memory microcontrollers and external memory microcontrollers, respectively.

Embedded memory microcontroller: A microcontroller having all the functional blocks on a single chip is called an embedded microcontroller. For example, the 8051 microcontroller contains program and data memory, input-output ports, serial communication port, counters, timers, and interrupts in one chip.

External memory microcontroller: A microcontroller not having all the functional blocks available on a single chip is called an external memory microcontroller. For example, the 8031 has no program memory on the chip; therefore, it is an external memory microcontroller.

(iii) *According to instruction set*:

The instructions or commands used in microcontroller programming are basically of two types: Reduced Instruction Set Computer (RISC) and Complex Instruction Set Computer (CISC).

RISC: It allows each instruction to operate on any register or use any addressing mode with simultaneous access of program and data. A RISC system reduces the execution time by decreasing the number of clock cycles per instruction.

CISC: It allows the programmer to use only one instruction in place of a number of simpler instructions. A CISC system reduces the execution time by decreasing the number of instructions per program.

(iv) *According to memory architecture*:

The process of instruction and data exchange between the memories inside the microcontroller depends on the architecture of memory mapping. Based on this architecture, microcontrollers are classified into two categories: Harvard memory architecture and Von Neumann memory architecture.

Harvard architecture: Microcontrollers based on this architecture have separate buses and different memory units for instructions and data. This allows simultaneous fetching of data and instructions as they are stored in different memory locations. Harvard memory architecture-based microcontrollers like PIC-microcontrollers are faster than Von Neumann memory architecture microcontrollers.

Von Neumann architecture: Microcontrollers based on this architecture share a common single bus and a common memory unit for instructions and data used during a program. In this type, the instructions are fetched before the data. Fetching of both data and instructions cannot occur simultaneously, e.g., memory in personal computers is of Von Neumann memory architecture.

4.2   How to Program a Microcontroller?

Microcontrollers are programmed either in an assembly language or in a high-level programming language like C, embedded C, C++, and python. To write code in a programming language, one needs a programming environment. This purpose is served by a software tool called *integrated development environment* (IDE). The IDEs are specific for a microcontroller family. An IDE consists of several built-in tools and libraries to program a microcontroller. The most common tools are a text editor, a compiler, and a debugger. Selecting a programming language for writing a code depends on the target microcontroller. Some IDEs have their own language for programming although they may also offer flexibility in choosing other languages, e.g., Arduino IDE.

- The first step to program a microcontroller is to write a code in text editor.
- Following this, the compiler translates it into a HEX code file, a format understandable by the microcontroller. The function of the debugger is to list the errors on compilation of the code, if any.
- After the generation of HEX file, it is uploaded to the microcontroller RAM using a programmer. Like an IDE, the programmer is also specific to a particular family of microcontrollers.
- Once the file is uploaded, the microcontroller, equipped with power supply and crystal frequency, is ready to be used for testing and operation. A schematic of the fundamental steps of programming a microcontroller is given in Fig. 4.2.

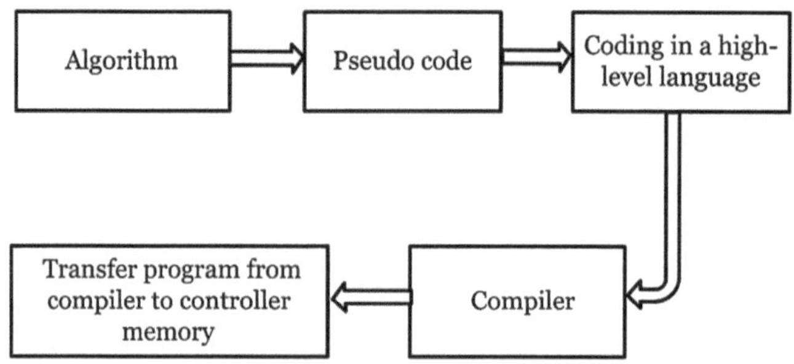

Fig. 4.2 Schematic of steps of programming a microcontroller

4.2.1   Factors for Selecting the Programming Language

Two of the important factors that may be considered for selecting a programming language to program a microcontroller are size and speed.

Size: The memory that the program occupies is very important as microcontrollers have limited memory.
Speed: The program execution is always desired to be fast in most of the applications. Programming languages with minimum number of machine cycles to decode are more preferable for faster operations.

One of the most commonly used programming languages for programming microcontrollers is Embedded C.

4.2.2   Some Common Concepts in Embedded C Programming

4.2.2.1   Data Types

Data types represent the nature of data to be used. Including the data types used for C programming, Embedded C uses a few more extra data types. The data types used in embedded C programming on a 32-bit machine are given in Table 4.1. The size and the range are different on machines with different word sizes.

4.2.2.2   Variables and Constants

Constants are values that remain fixed during the execution of a program. Variables are names assigned to memory addresses to store constants. Unlike constants, variables can be changed during the execution of a program. A variable has to belong to a specific data type for knowing the type of data it holds.

Table 4.1 Data types used in Embedded C

Type	Number of bits (size)	Range of value
int or signed int	16	−32768 to +32767
unsigned int	16	0 to 65535
char or signed char	8	−128 to 127
unsigned char	8	0 to 255
double	32	±1.175494E-38 to ±3.402823E+38
float	32	±1.175494E-38 to ±3.402823E+38
bit	1	0 or 1
sbit	1	0 or 1 (bit addressable part of RAM)
sfr	8	RAM addresses (80H TO FFH)
sfr16	16	0 to 65535

4.2.2.3 Keywords

Keywords are the words that convey special meanings to the language compiler. They are reserved words for special purpose and are predefined and standard in nature. They always begin with a lowercase.

The basic keywords of an embedded software are sbit, bit, and sfr.

sbit: This data type is used in case of accessing a single bit of SFR register.

bit: This data type is used for accessing the bit addressable memory of RAM (20h-2fh).

sfr: This data type is used for accessing an SFR register by another name. All the SFR registers must be declared with capital letters.

4.2.3 The Structure of an Embedded C Program

The difference between Embedded C and generic C program is tabulated in Table 4.2.

The structure of an embedded C program consists of the following sections:

 (i) Documentation
(ii) Pre-processor directives
(iii) Global variables declaration

Table 4.2 A comparison of Embedded C and generic C program

Generic C program	Embedded C
Used for developing general-purpose applications on desktop or server environments	Used for programming embedded systems that perform dedicated functions
Generally has access to abundant resources, such as large amounts of memory and powerful processors	Operates within constrained environments with limited memory, processing power, and other resources
Uses standard libraries provided by the operating system or development environment	Uses hardware-specific libraries and direct access to hardware registers
Code is generally portable across different platforms and operating systems if the standard C libraries are available	Code is usually highly specific to the hardware it is designed for
Developed using general-purpose IDEs and compilers, such as GCC, Visual Studio, or Clang	Uses specialized development tools tailored for embedded systems, such as AVR Studio, Keil uVision, MPLAB, and hardware-specific compilers
Limited direct interaction with hardware; usually relies on operating system abstractions for hardware access	Directly interacts with hardware components
Generally, relies on dynamic memory allocation	Uses static memory allocation

(iv) main ( ) function

```
{
local variable declaration
statements
}
```

(v) Subprogram section

```
function1 ( )
{
local variables declaration
statements
}
```

```
function2 ( )
{
local variables declaration
statements
}
```

4.2.3.1   Documentation/Comments

Documentation is an important part of any code. It consists of a set of comment lines which help the reader to understand the code easily. These texts are ignored by the compiler. It includes single line and multiline comments. A part of documentation before the beginning of a code is given below:

```
/* multi line comments
@Author list: XYZ
@Filename: abc
@Functions: delay (int sec), main ()
*/
#define port2 P2 //single line comment
```

4.2.3.2   Pre-processor Directives

A pre-processor directive in embedded C is an indication to the compiler that it must look into this file for symbols that are not defined in the program. Usually in Embedded C, the pre-processor directive is used to indicate a header file specific to the microcontroller, which contains all the SFRs and the bits in those SFRs.

The pre-processor directives for programming an 8051 microcontroller in Keil uVision compiler are *#include* and *#define.*

In Keil μVision IDE, the pre-processor directives are written as

```
#include <reg51.h> //special function register 8051
#define port2 P2 //declaring port
```

4.2.3.3   Global Variable

A global variable is a variable that can be accessed by more than one function and can be defined anywhere in a program. These are declared outside the function.

For instance, in Keil μ-vision IDE, the global variables are declared as follows:

```
sbit led=port 1^0;//global declaration
unsigned int p; //global variable declaration
```

4.2.3.4   Local Variable

Local variable is a variable which is declared inside a function and can be accessed by that function only.

The code snippet here shows the declaration of a local variable within a function

```
void delay (unsigned int msec) {
int i, j ;// local variable declaration
for (i=0; i<msec; i++)
for (j=0; j<1275; j++)
}
```

4.2.4   Main () function

The main function is the function from which the execution of a program begins. The program execution begins at the opening brace and ends at the closing brace.

(i) void main(void): The void main(void) tells that the main () will not return any value.

For example, the code snippet for interfacing LED with 8051 in Keil μ-vision is given below

```
Main (void)
{ //opening braces
While (1)
{
delay (100);
led = 0;
delay (100);
led = 1;
}
} //closing braces
```

4.2.5   Subprogram Section

This includes the user-defined functions that are called in the main function, e.g., the delay function is a user-defined function which can be expressed as

```
void delay (unsigned int msec)
{
int i, j ;// local variable declaration
for (i=0; i<msec; i++) //for loop
for (j=0; j<1275; j++)
}
```

4.2.6   An Example of the Embedded C Programming

```
. . . . . . . . . . . . . . . . . . . . . . . . . . . . .
DOCUMENTATION. . . . . . . . . . . . . . . . . . . . . . . . . . . .
/*
Project name: LED interfacing with 8051 microcontroller
Author list: XYZ
Filename: led_blink. uvproj
Functions: delay (unsigned int msec), main ()
*/
. . . . . . . . . . . . . . . . . . . . . . . . . . . . . . . . . . . . . . . . . . . . . .
. . . . . . . . . . . . . . . . . . . . . . . .
#include <reg51.h> //PREPROCESSOR DIRECTIVE
#define port1 P1 //port declaration
. . . . . . . . . . . . . . . . . . . . . . . . . . . . . .GLOBAL
Variables. . . . . . . . . . . . . . . . . . . . . . . . . . .
sbit led=port1^0; //global declaration
unsigned int msec;
```

```
......................FUNCTION
Declaration.............................
void delay ();
.......................MAIN
FUNCTION...............................
int main (void)
{//opening braces
while (1) //
{
delay (100);
led = 0;// led off
delay (100);
led = 1;//led on
}
} //closing braces


.....................SUBPROGRAM
SECTION..............................
void delay (unsigned int msec)
{
int i, j ;// local variable declaration
for (i=0; i<msec; i++) //for loop
for (j=0; j<1275; j++)
}
```

4.3   Commonly Used Microcontrollers in Robotics

4.3.1   8051

The 8051 microcontroller, also known as Intel MCS-51, is a single integrated chip microcontroller (MCU) series launched by Intel in 1981. It is an 8-bit microcontroller. The 8051 microcontroller is the most popularly used microcontroller. The 8051 architecture provides many functions through its central processing unit (CPU), random access memory (RAM), read-only memory (ROM), input/output (I/O) ports, serial port, interrupt control, and timers in one package. AT89C51 and AT89S52 are examples of commercial 8051 microcontroller.

4.3.1.1   AT89C51

The AT89C51 is a variant of the original Intel 8051 8-bit microcontroller from the Atmel89 series family. It works with the popular 8051 architecture and is the mostly used microcontroller till date. It is a popular microcontroller to begin learning on embedded systems. The original 8051 was developed using N-type metal oxide semiconductor (MOS) technology, whereas AT89C51 was developed using CMOS technology because of its low power utilization.

It is a 40-pin IC package with 4-Kb flash programmable and erasable read-only memory (PEROM). It has four ports and all together provides 32 programmable GPIO pins. It does not have in-built ADC module and supports only USART communication. However, it can be interfaced with external ADC IC like the ADC084 or the ADC0808.

The device is manufactured using Atmel high-density non-volatile memory technology and is compatible with the industry-standard MCS-51 instruction set and pinouts. The on-chip flash allows the program memory to be reprogrammed in-system or by a conventional non-volatile memory programmer. By combining a versatile 8-bit CPU with flash on a monolithic chip, the Atmel AT89C51 is a powerful microcomputer, which provides a highly flexible and cost-effective solution to many embedded control applications.

- Pin diagram (Fig. 4.3)

Programming the AT89C51

Atmel microcontrollers can be programmed with different software available in the market. Arduino and Keil uVision are the most used platforms out of which Keil uVision is used most widely. In order to program the Atmel microcontroller, one needs an integrated development environment (IDE), where the programming takes place. A compiler, where the program gets converted into MCU readable form, is called a hex files. An integrated programming environment (IPE) is used to dump the hex file into the MCUs.

Fig. 4.3 Pin diagram of AT89C51

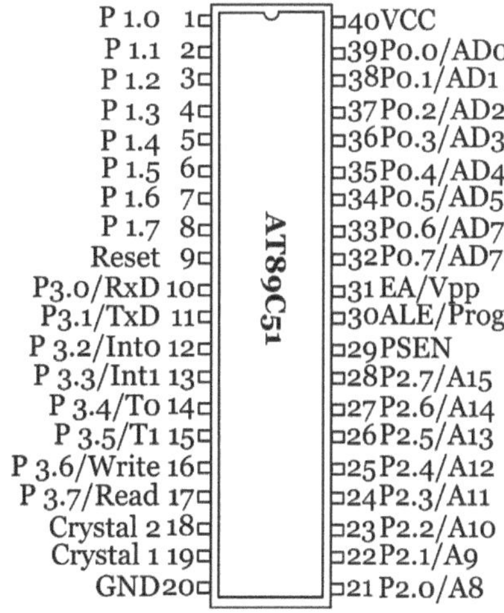

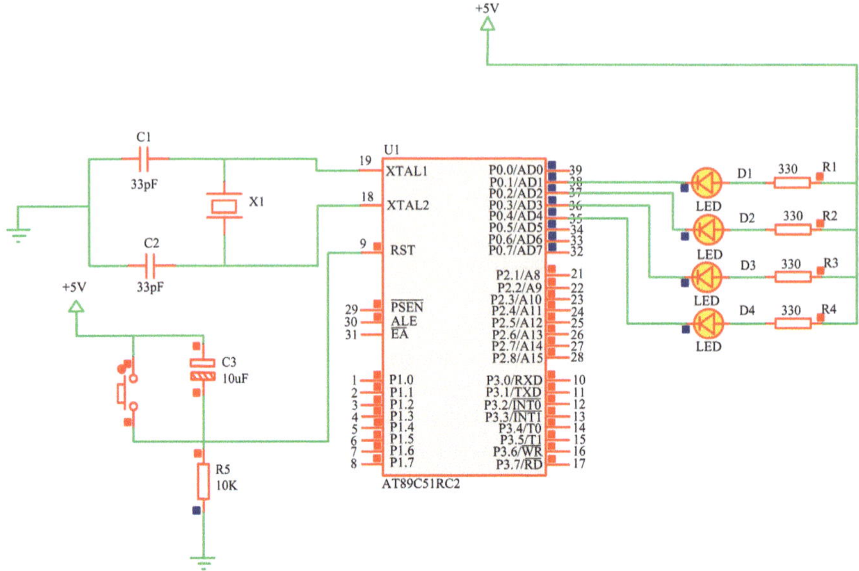

Fig. 4.4 Schematic showing LED interface with AT89C51

To dump or upload the code into Atmel IC, one needs a programmer. The most commonly used programmer is the USB ASP which has to be purchased separately. Also simulating the program on a software like Proteus before trying it on hardware saves time.

Example Program: Interfacing LED with AT89C51 (Fig. 4.4)

4.3.2 PIC Microcontroller

The PIC microcontroller was developed in the year 1993 by microchip technology. The term PIC stands for Peripheral Interface Controller. Initially, this was developed for supporting programmable data processor (PDP) computers to control their peripheral devices and, therefore, was named as a peripheral interface device. These microcontrollers are of high speed and execution of a program is easy as compared to other microcontrollers.

PIC microcontrollers are the world's smallest microcontrollers that can be programmed to carry out a huge range of tasks. These microcontrollers are found in many electronic devices such as phones, computer control systems, alarm systems, and embedded systems. Various types of microcontrollers exist, even though the best is found in the GENIE range of programmable microcontrollers.

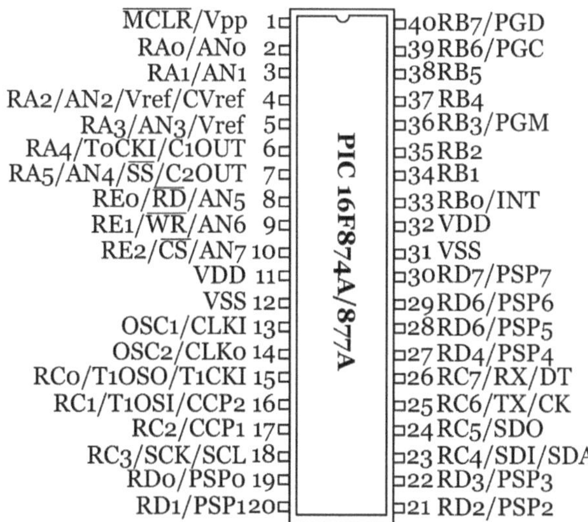

Fig. 4.5 Pin diagram of PIC 16F874A/877A

Every PIC microcontroller architecture consists of some registers and stack where the registers function as Random Access Memory (RAM) and the stack saves the return addresses. The main features of PIC microcontrollers are RAM, flash memory, Timers/Counters, EEPROM, I/O Ports, USART, CCP (Capture/Compare/PWM module), SSP, Comparator, ADC (analog to digital converter), PSP (parallel slave port), LCD, and ICSP (in circuit serial programming). The 8-bit PIC microcontroller is classified into four types on the basis of internal architecture: base line PIC, mid-range PIC, enhanced mid-range PIC, and PIC18.

Pin Diagram (Fig. 4.5):

4.3.2.1   Programming the PIC Microcontroller

PIC microcontrollers can be programmed with different softwares available in the market. There is still use of assembly language to program PIC MCUs. The below-mentioned details are for the most advanced and common software and compiler that has been developed by Microchip itself. In order to program the PIC microcontroller, one need an Integrated Development Environment (IDE), where the programming takes place; a compiler, where the program gets converted into MCU readable form called HEX files; and an Integrated Programming Environment (IPE), which is used to dump our hex file into our PIC MCUs.

To dump or upload the code into PIC, one will need a device called PICkit3. The PICkit3 programmer/debugger is a simple, low-cost in-circuit debugger that is controlled by a PC running MPLAB IDE (v8.20 or greater) software on a Windows platform. The PICkit3 programmer/debugger is an integral part of the development engineer's tool suite. In addition to this, one will also need other hardware like Perf board or breadboard, soldering station, PIC ICs, crystal oscillators, and capacitors.

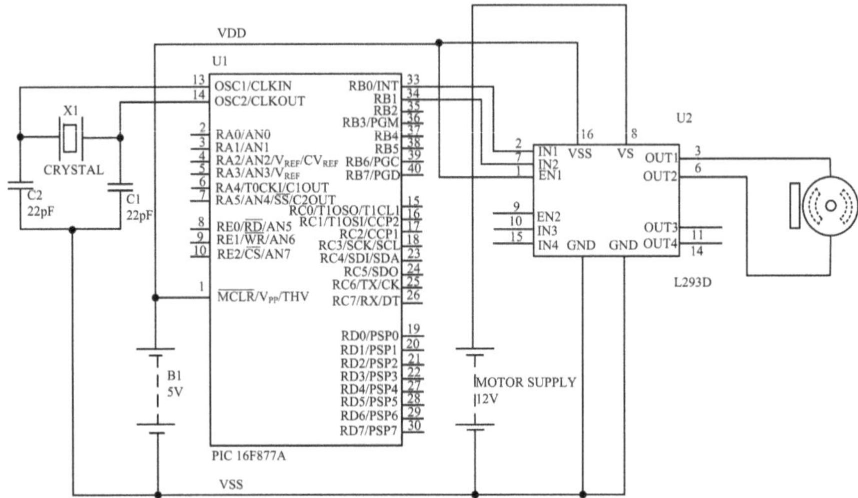

Fig. 4.6 Schematic of interfacing a DC motor with PIC microcontroller

4.3.2.2   Example Program: Interfacing DC Motor with PIC Microcontroller (Fig. 4.6)

```
void main ()
{
TRISB = 0; // PORT B as output port
PORTB = 1; // Set RB0 to high
do
{
//To turn motor clockwise
PORTB.F0 = 1;
Delay_ms (2000); //2 second delay

//To Stop motor
PORTB = 0; // or PORTB = 3
Delay_ms (2000); //2 second delay

//To turn motor anticlockwise direction
PORTB.F1 = 1;
Delay_ms(2000);//2 second delay
```

```
//To Stop motor
 PORTB = 0; // or PORTB = 3 (3 = 0b00000011)
Delay_ms(2000); // 2 seconds delay

} while (1);
}
```

4.3.3   ATMEGA Microcontroller

ATMega microcontrollers belong to the AVR family of microcontrollers and are manufactured by Atmel Corporation. An ATMega microcontroller is an 8-bit microcontroller with Reduced Instruction Set (RISC)-based Harvard Architecture. It has standard features like on-chip ROM (Read-Only Memory), Data RAM (Random Access Memory), data EEPROM (Electrical Erasable Programmable Read-Only Memory), Timers and Input/Output Ports, along with additional peripherals like ADC and Serial Interface Ports.

4.4   LED Blink Using an Arduino

This example demonstrates an LED blink using an Arduino. To build this circuit, one end of the resistor is connected to the digital pin corresponding to the LED_BUILTIN constant. The positive leg of the LED is connected to the other end of the resistor. The negative leg of the LED is connected to the ground. Figure 4.7 illustrates an UNO board where D13 serves as the LED_BUILTIN value.

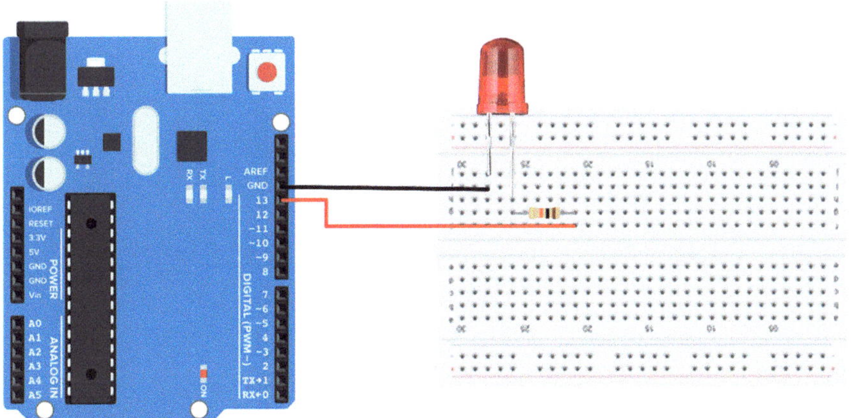

Fig. 4.7 Circuit diagram for LED blink using an Arduino

After building the circuit, the Arduino board should be plugged into the computer. The Arduino Software (IDE) is then started, and the code below is entered. The first step is to initialize the LED_BUILTIN pin as an output pin with the line

```
pinMode(LED_BUILTIN, OUTPUT);
```

In the main loop, you turn the LED on with the line:

```
digitalWrite(LED_BUILTIN, HIGH);
```

This supplies 5 volts to the LED and lights it up. Then you turn it off with the line:

```
digitalWrite(LED_BUILTIN, LOW);
```

That takes the LED_BUILTIN pin back to 0 volts, turning the LED off. In between the on and off states, there should be enough time for a person to see the change, so the delay() commands instruct the board to do nothing for 1000 milliseconds.

4.4.1 Code

```
void setup()
{
pinMode(LED_BUILTIN, OUTPUT); // initialize digital pin as an output
}
void loop() {
digitalWrite(LED_BUILTIN, HIGH); // turn the LED on
delay(1000); // wait for a second
digitalWrite(LED_BUILTIN, LOW); // turn the LED off
delay(1000); // wait for a second
}
```

Chapter 5
Spatial Descriptions: Frames and Transformations

It is absolutely important for an automated robotic system to correctly manipulate its joints and links. To command and control such manipulation in the form of electronic signals, it is essential to describe the desired position and the orientation of different robotic parts. This necessitates the purpose for defining a coordinate system and its transformations.

5.1 Coordinate System

Cartesian coordinate system: Cartesian coordinate system is used to specify the position and the orientation or the pose of any point or an object in a three-dimensional space. The position is defined by perpendicular projection of the point or object onto three mutually perpendicular lines termed as axes of a coordinate system. The orientation of an object is defined by angles of the object with these three perpendicular lines.

Figure 5.1a shows position of a point P (x, y, z) in a three-dimensional space with its projections P_x, P_y, P_z onto the X, Y, Z axes, respectively. Figure 5.1b shows the position and the orientation of an object in three-dimensional space with its projection P_x, P_y, P_z onto the X, Y, Z axes, respectively, and α, β, and γ angles with the X, Y, Z axes, respectively. The coordinate system with X, Y, Z axes is called frame {X, Y, Z}.

Polar coordinate system: Polar coordinate system in three-dimensional space determines the position of a point or an object by a distance from a reference point or origin of the frame and an angle from a reference direction. Figure 5.1c shows a point P in polar coordinate system which is represented by three numbers: (a) radial distance ρ of the point from origin of the frame, (b) polar angle θ measured from a fixed zenith direction, and (c) azimuthal angle φ of its orthogonal projection onto the reference plane that passes through the origin.

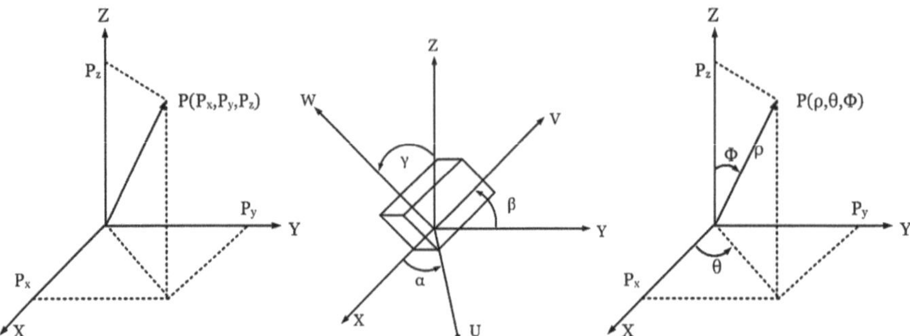

Fig. 5.1 (**a**) Cartesian coordinate system; (**b**) orientation of an object expressed using angles α, β, and γ of the object with the coordinate axes; (**c**) polar coordinate system showing the polar coordinates (ρ, θ, φ) of a point P

5.1.1 Position and Orientation in a Coordinate System

In the study of kinematics of a robotic system, one has to deal with the position and the orientation of several bodies in space. To describe the position and orientation or pose of a body in a coordinate system, two coordinate frames are used: fixed or global frame, and moving or local frame.

5.1.1.1 Fixed and Moving Frame

The position and orientation of a rigid body in a three-dimensional space are represented with respect to a fixed reference coordinate system in the space. This coordinate frame is known as the fixed or global frame.

The coordinate system attached to the moving bodies used to express their pose with respect to the fixed frame in space is known as the moving or local frame. In Fig. 5.2, the schematic of a fixed frame and a moving frame is shown.

5.2 Coordinate Transformation

Translational and rotational transformations are two types of transformations associated with a body in space. The translation brings changes in the location of the body, whereas the rotation brings changes in its orientation. The translation and rotation of an object are expressed by co-locating the moving frames attached to the body with respect to the fixed frame.

These transformations are expressed in terms of matrices representing the translation and rotation of a body. The matrix representing the translation of an object or moving frame with reference to a fixed frame is called translation matrix. The matrix

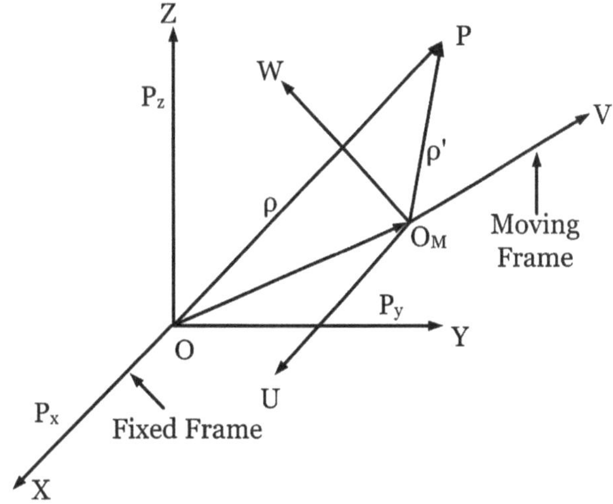

Fig. 5.2 Spatial description of fixed and moving frame. {X, Y, Z} is the fixed frame with its origin at O. {U, V, W} is moving frames with origins at O_M. The pose of any bodies attached to the moving frames {U, V, W} is described by describing the pose of the moving frames with respect to the fixed frame {X, Y, Z}

representing the rotation of an object or moving frame with respect to a fixed frame is called rotational matrix.

Derivation of the translation and rotational matrices is beyond the scope of this book and readers can refer to Chapter 5 of the book entitled "Introduction to Robotics" by Subir Kumar Saha for the purpose. For further discussions, we will be considering the translational and the rotational matrices in their final forms.

5.2.1 Translational Transformation

Translation of an object in a coordinate frame {X, Y, Z} can happen along the X, Y, and Z axes. Considering the vectors P_x, P_y, and P_z representing the translation of the object along the x, y, and z axes, respectively, the translation matrix representing this translation is expressed as (Fig. 5.3):

$$T = \begin{bmatrix} P_x \\ P_y \\ P_z \end{bmatrix}$$

5.2.2 Rotational Transformation

Rotation of an object in a coordinate frame {X, Y, Z} can happen about the X, Y, and Z axes. The matrices R_x, R_y, and R_z representing the rotation about X, Y, and Z axes respectively are (Fig. 5.4)

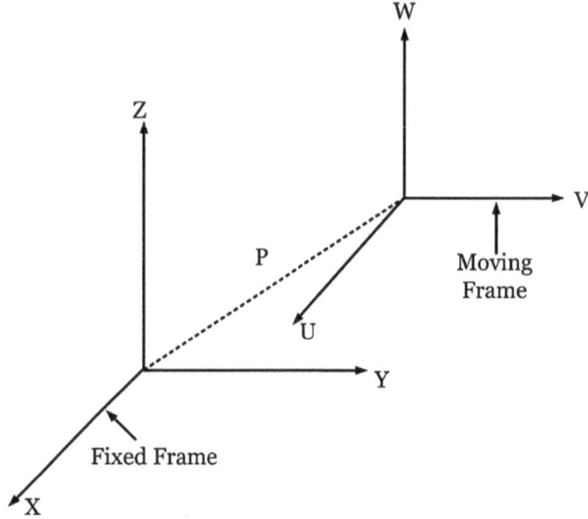

Fig. 5.3 Translation of a moving object attached to a frame {U, V, W} with respect to a fixed frame along the Y and Z axes. Translation of a fixed frame {X, Y, Z} along the Y and Z axes to reach the moving frame {U, V, W}

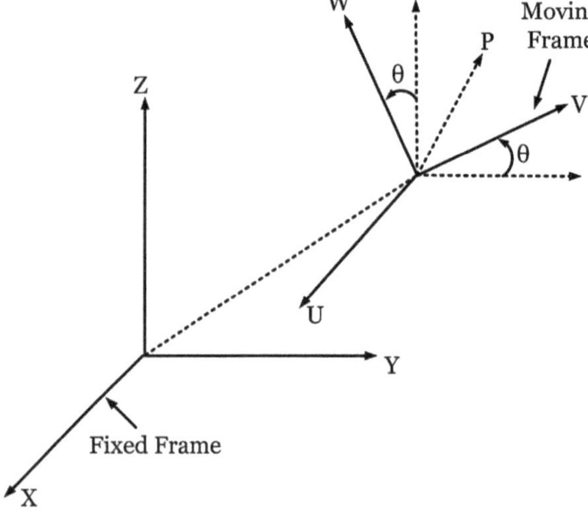

Fig. 5.4 Rotation of a moving body attached to the moving frame {U, V, W} by an angle θ about the X axis with respect to a fixed frame {X, Y, Z}. Rotation of a fixed frame {X, Y, Z} by an angle θ about the X axis to reach the moving frame {U, V, W}

$$R_x = \begin{bmatrix} 1 & 0 & 0 \\ 0 & \text{Cos}\gamma & -\text{Sin}\gamma \\ 0 & \text{Sin}\gamma & \text{Cos}\gamma \end{bmatrix}, R_y = \begin{bmatrix} \text{Cos}\beta & 0 & \text{Sin}\beta \\ 0 & 1 & 0 \\ -\text{Sin}\beta & 0 & \text{Cos}\beta \end{bmatrix}, R_z = \begin{bmatrix} \text{Cos}\alpha & -\text{Sin}\alpha & 0 \\ \text{Sin}\alpha & \text{Cos}\alpha & 0 \\ 0 & 0 & 1 \end{bmatrix}$$

5.3   Homogeneous Transformation Matrix

A homogeneous transformation matrix combines a translational and a rotational transformation matrix into one 4×4 matrix.

$$H = \begin{bmatrix} R & T \\ O_{[1 \times 3]} & 1 \end{bmatrix}$$

R is a rotational matrix that depends on the rotation about the X, Y, or Z axis. T is the translational matrix and $\mathbf{0}$ is a 1×3 matrix.

Considering rotation of a fixed frame about Z axis and its translation along the Y and Z axes, the homogenous matrix is written as:

$$H = \begin{bmatrix} \cos\alpha & -\sin\alpha & 0 & P_x \\ \sin\alpha & \cos\alpha & 0 & P_y \\ 0 & 0 & 1 & P_z \\ 0 & 0 & 0 & 1 \end{bmatrix}$$

5.4   Denavit Hertenberg Parameters

The Denavit Hertenberg (DH) parameters is a set of four parameters used to describe the position and orientation of a link or a joint with respect to its previous or succeeding link or joint. To control the pose of a manipulator's end effector for performing specific tasks, the position and orientation of the end effector need to be expressed with respect to the manipulator's base. It can be done by describing the transformation of the coordinate frames attached to end-effector to the base through the intermediate joints. The description of the transformation can be easily accomplished using the DH parameters: joint offset, joint angle, link length, and twist angle.

Figure 5.5 is drawn to define these parameters. The transformation between the Frame *i+1* attached to Link *i* and the Frame *i* attached to Link *i-1* is represented using the four DH parameters where *i* is the index of the link. A coordinate frame is attached to each link.

With respect to Fig. 5.5, the DH parameters are defined as:

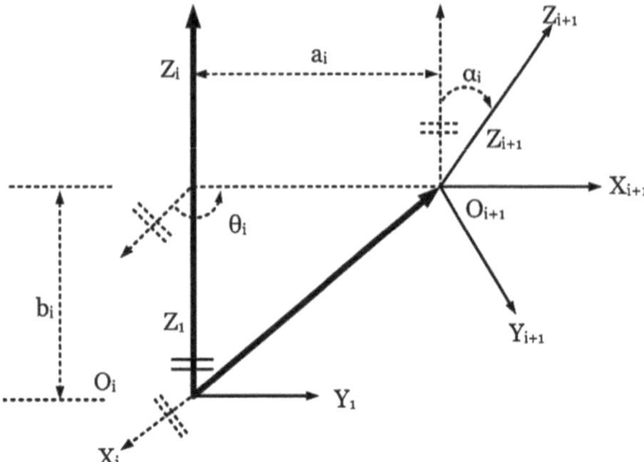

Fig. 5.5 Schematic of DH parameters using two successive coordinate frames

(a) Joint offset (b_i): Distance between X_i and X_{i+1} along Z_i
(b) Joint angle (θ_i): Angle between X_i and X_{i+1} about Z_i
(c) Link length (a_i): Distance between Z_i and Z_{i+1} along X_{i+1}
(d) Twist angle (α_i): Angle between Z_i and Z_{i+1} about X_{i+1}

Chapter 6
Kinematics and Dynamics

6.1 Introduction

Kinematics is the study of motion of a robotic system without considering the cause of motion. There must be force or torque causing a linear or rotary motion. In kinematics, one does not try to find what the amount of force or torque should be. Kinematics deals with the motion of the system and the relative motion between different links.

To study the motion of a robotic system, it is required to establish the orientation or pose (i.e., the information regarding position and orientation) of the links with respect to its previous link (or base). For this purpose, the Cartesian coordinates of the end-effector (position and orientation of a point on the end-effector) need to be determined. The relation of pose between the successive links needs to be determined. Based on analysis, there are two types of problems in kinematics:

- Forward or direct kinematics (forward position analysis)
- Inverse kinematics (inverse position analysis)

If the joint angles are given for a particular robotic manipulator and one needs to find the pose of end-effector, it is called forward position analysis. If the pose of the end-effector is given or known and one needs to determine the joint angles, it is called inverse position analysis. A forward position analysis always has a fixed single solution. But inverse position analysis may have more than one solution. The pictorial representation of forward and inverse kinematics is shown in Fig. 6.1.

Two main solution techniques for the inverse kinematics problem are analytical and numerical methods. In the first type, the joint variables are solved analytically according to given configuration data. In the second type of solution, the joint variables are obtained based on numerical techniques.

There are two approaches in analytical methods: geometric and algebraic approaches. Geometric approach is applied to simple robot structures, such as 2-DoF planar manipulator or less DoF manipulator with parallel joint axes.

© The Author(s), under exclusive license to Springer Nature Switzerland AG 2024
N. M. Kakoty et al., *Introduction to Embedded Systems and Robotics*,
https://doi.org/10.1007/978-3-031-73098-6_6

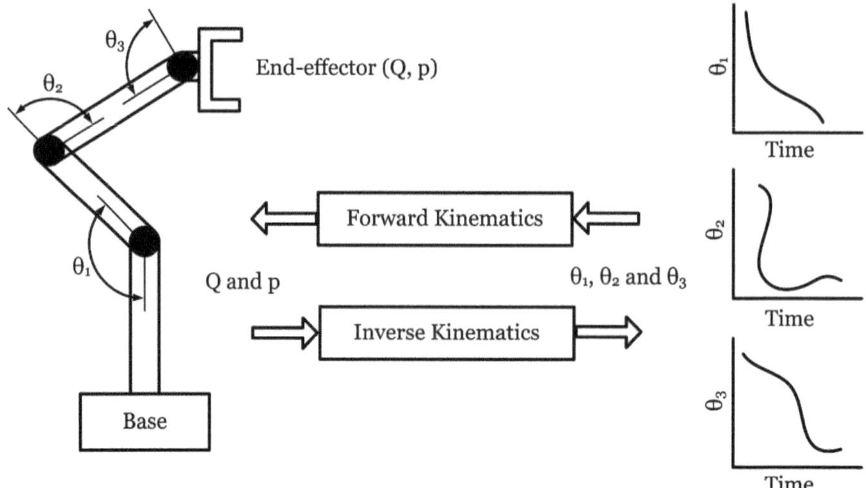

Fig. 6.1   Schematic of forward and inverse kinematics

Manipulators with more links with arms extended into three dimensions or geometry become more tedious. In such cases, algebraic approach is more beneficial for inverse kinematics solution.

6.2   Forward Position Analysis

In the forward position analysis, the joint angles (joint displacement in case of prismatic joints) are available. Based on these values, one needs to find the position and orientation of the end-effector.

First, let us discuss about robot mechanisms that work within a plane, i.e., planar kinematics. Let us consider the three DoF planar robot arm shown in Fig. 6.2. The arm consists of one fixed link and three movable links that move within the plane. All the links are connected by revolute joints whose joint axes are all perpendicular to the plane of the links. There is no closed-loop kinematic chain and, therefore, it is a serial link mechanism.

One can relate the end-effecter coordinates to the joint angles and link lengths as:

$$X_e = l_1 \cos\theta_1 + l_2 \cos(\theta_1 + \theta_2) + l_3 \cos(\theta_1 + \theta_2 + \theta_3) \tag{6.1}$$

$$Y_e = l_1 \sin\theta_1 + l_2 \sin(\theta_1 + \theta_2) + l_3 \sin(\theta_1 + \theta_2 + \theta_3) \tag{6.2}$$

The orientation of the end-effecter can also be described as the angle made by the center-line of the end-effector with the positive X-coordinate axis. This end-effector orientation Q_e is related to the joint angles as:

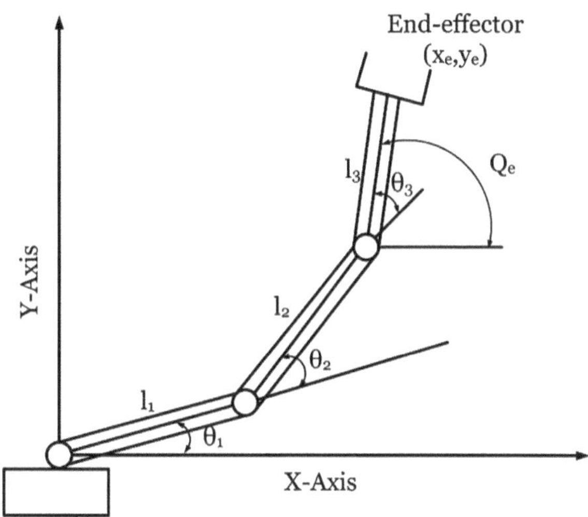

Fig. 6.2 Three DoF planar robot with three revolute joints

$$Q_e = \theta_1 + \theta_2 + \theta_3 \tag{6.3}$$

The above three equations describe the position and orientation of the robot end-effector as viewed from the fixed coordinate system in relation to the joint angles. In general, a set of algebraic equations relating the position and orientation of a robot end-effector or any significant part of the robot to the joint angles is called kinematic equations or, more specifically, forward kinematic equations in robotics.

Planar kinematics is mathematically much more tractable compared to three-dimensional kinematics. For three-dimensional forward position analysis, one can follow the four standard steps given below.

Step 1: Attach the coordinate frames in each of the links.
Step 2: Define the Denavit Hartenberg (DH) parameter for every link.
Step 3: Write the homogeneous transformation matrix (HTM) of each frame with respect to the previous frame.
Step 4: The resultant HTM of end-effector with respect to the base is determined by post-multiplication of the previous individual HTMs.

To understand the above steps, one can consider the following serial manipulator in Fig. 6.3 and perform the forward position analysis.

Attach coordinate frames to each of the n + 1 links of the robot, with frame 1 attached to the fixed link and frame n+1 attached to the end-effector. Define the DH parameters for every link, i.e., link #1, link #2, link #3,................., link #n. The HTMs T_1, T_2, T_3, ...T_n, where T_i for i = 0, 1, 2, ..., n represents the transformation of body i with respect to its previous link i-1 or frame i+1 with respect to the frame attached to it, i.e., frame i.

Now, obtain the individual HTM from the four elementary transformations corresponding to the DH parameters. The HTM of the end-effector (frame n + 1)

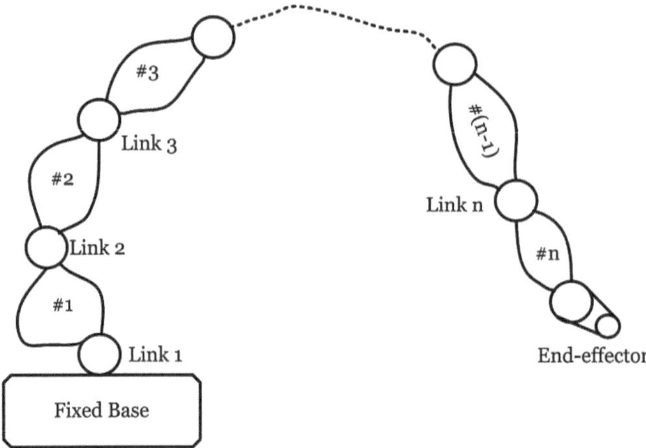

Fig. 6.3 Schematic of a serial manipulator

with respect to frame 1, i.e., T, is obtained by post-multiplication of the previous individual HTM, i.e., T_i for i = 0, 1, 2, ..., n.

This can be expressed as:

$$T = T_1 \, T_2 \ldots \ldots T_n \tag{6.4}$$

Equation (6.4) is called the closure equation of the robot. Orientation Q of the end-effector with respect to the fixed frame can be expressed as:

$$Q = Q_1 \, Q_2 \ldots \ldots Q_n \tag{6.5}$$

where, Q_i is orthogonal rotation matrix used to represent orientation of frame i+1 with respect to frame i.

Example 6.1 This example is to understand how to determine the configuration of end-effector of the two-link planar arm shown in Fig. 6.4 using forward position analysis.

To proceed, one needs to attach the coordinate frame as shown in Fig. 6.5 as the first step.

The next step is to find the DH parameter of the two-link arm as in Table 6.1.

HTM of the links is given by the matrix:

$$T_i = \begin{bmatrix} c_i & -s_i & 0 & a_i c_i \\ s_i & c_i & 0 & a_i s_i \\ 0 & 0 & 1 & 0 \\ 0 & 0 & 0 & 1 \end{bmatrix}$$

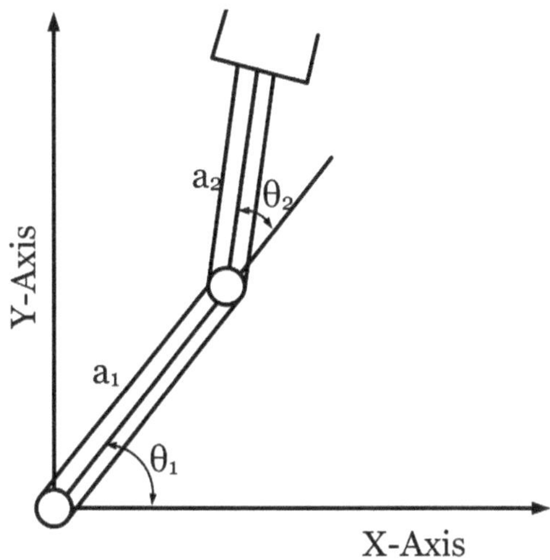

Fig. 6.4 Schematic of a two-link planner arm

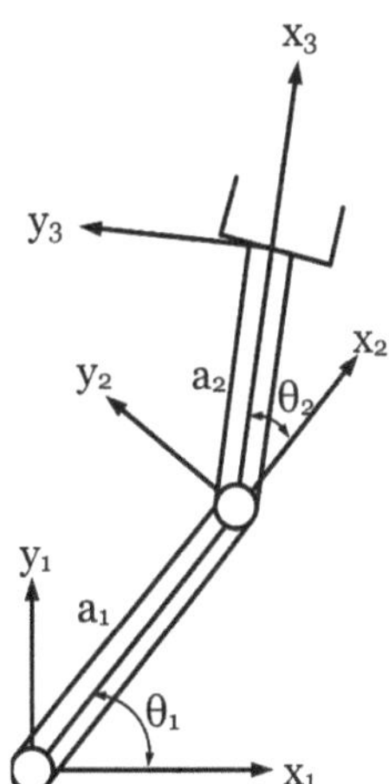

Fig. 6.5 Two-link planner arm with coordinate frame

Table 6.1 DH parameter for the two-link arm in Fig. 6.5

Link	b_i	θ_i	a_i	α_i
1	0	θ_1	a_1	0
2	0	θ_2	a_2	0

where, T_i represents the HTM for links 1 and 2 having $c_i = \cos\theta_i$, $s_i = \sin\theta_i$, $a_i =$ link length with $i = 1, 2$.

The solution of forward position analysis is:

$$T = T_1 . T_2$$

$$T = \begin{bmatrix} c_1 & -s_1 & 0 & a_1 c_1 \\ s_1 & c_1 & 0 & a_1 s_1 \\ 0 & 0 & 1 & 0 \\ 0 & 0 & 0 & 1 \end{bmatrix} \cdot \begin{bmatrix} c_2 & -s_2 & 0 & a_2 c_2 \\ s_2 & c_2 & 0 & a_2 s_2 \\ 0 & 0 & 1 & 0 \\ 0 & 0 & 0 & 1 \end{bmatrix}$$

$$= \begin{bmatrix} c_{12} & -s_{12} & 0 & a_1 c_1 + a_2 c_2 \\ s_{12} & c_{12} & 0 & a_1 s_1 + a_2 s_2 \\ 0 & 0 & 1 & 0 \\ 0 & 0 & 0 & 1 \end{bmatrix}$$

where,

$$\left. \begin{array}{l} c_{12} = \cos\theta_{12} \\ s_{12} = \sin\theta_{12} \end{array} \right\} \quad \theta_{12} \equiv \theta_1 + \theta_2$$

$$P_x = a_1 c_1 + a_2 c_2 \tag{6.6}$$

$$P_y = a_1 s_1 + a_2 s_2 \tag{6.7}$$

By knowing the values of link lengths (a_1 and a_2) and joint angles (θ_1 and θ_2), one can determine the configuration of end-effector of the two-link planar arm.

6.3   Inverse Kinematics

The problem of finding the set of joint angles for a given end-effector position and orientation is referred to as inverse kinematics or inverse position analysis. In this section, the problem of moving the end-effector of a manipulator arm to a specified position and orientation is discussed. For this, one needs to find the joint angles that lead the end-effector to the specified position and orientation. This is the inverse of the previous problem and is thus referred to as inverse kinematics. The kinematic equation must be solved for joint angles, given the end-effecter position and orientation. Once the kinematic equation is solved, the desired end-effector motion can be achieved by moving each joint to the determined pose.

In forward kinematics problem, the end-effecter location is determined uniquely for any given set of joint angles. On the other hand, inverse kinematics is more complex in the sense that multiple solutions may exist for the same end-effecter pose. Also, solutions may not always exist for a particular range of end-effector poses and joint angles. Furthermore, since the kinematic equation is comprised of nonlinear equations with trigonometric functions, it is not always possible to derive a closed-form solution, which is an explicit inverse function of the kinematic equation. There are two solution approaches: geometric and algebraic used for deriving the analytical inverse kinematics solutions. When the kinematic equation cannot be solved analytically, numerical methods are used in order to derive the desired joint angles.

6.3.1   Geometric Solution Approach

Geometric solution approach is based on decomposing the spatial geometry of the manipulator into plane geometry problems. Application of this approach to a 2-DoF planar manipulator with revolute joints and link lengths l_1 and l_2 as shown in Fig. 6.6 is considered for discussion.

The components of the point P (p_x, p_y) are determined as:

$$p_x = l_1 c\theta_1 + l_2\, c\theta_{12};\ \text{where}\ c\theta_{12} = c\theta_1.c\theta_2 - s\theta_1.s\theta_2 \tag{6.8}$$

$$p_y = l_1 s\theta_1 + l_2\, s\theta_{12};\ \text{where}\ s\theta_{12} = s\theta_1.c\theta_2 + c\theta_1.s\theta_2 \tag{6.9}$$

Squaring both sides of Eqs. (6.8) and (6.9),

$$p_x^2 = l_1^2 c^2\theta_1 + l_2^2 c^2\theta_{12} + 2l_1 l_2\, c\theta_1.c\theta_{12} \tag{6.10}$$

$$p_y^2 = l_1^2 s^2\theta_1 + l_2^2 s^2\theta_{12} + 2l_1 l_2\, s\theta_1.s\theta_{12} \tag{6.11}$$

Using Eqs. (6.10) and (6.11),

$$\begin{aligned}
p_x^2 + p_y^2 &= l_1^2(c^2\theta_1 + s^2\theta_1) + l_2^2(c^2\theta_{12} + s^2\theta_{12}) + 2l_1 l_2\, (c\theta_1.c\theta_{12} + s\theta_1.s\theta_{12}) \\
&= l_1^2 + l_2^2 + 2l_1 l_2\, \{c\theta_1.(c\theta_1.c\theta_2 - s\theta_1.s\theta_2) + s\theta_1.(s\theta_1.c\theta_2 + c\theta_1.s\theta_2)\} \\
&= l_1^2 + l_2^2 + 2l_1 l_2\, \{c^2\theta_1.c\theta_2 - c\theta_1.s\theta_1.s\theta_2 + s^2\theta_1.c\theta_2 + s\theta_1.c\theta_1.s\theta_2\} \\
&= l_1^2 + l_2^2 + 2l_1 l_2\{c\theta_2\,(c^2\theta_1 + s^2\theta_1)\} \\
&= l_1^2 + l_2^2 + 2l_1 l_2 c\theta_2
\end{aligned}$$

$$\tag{6.12}$$

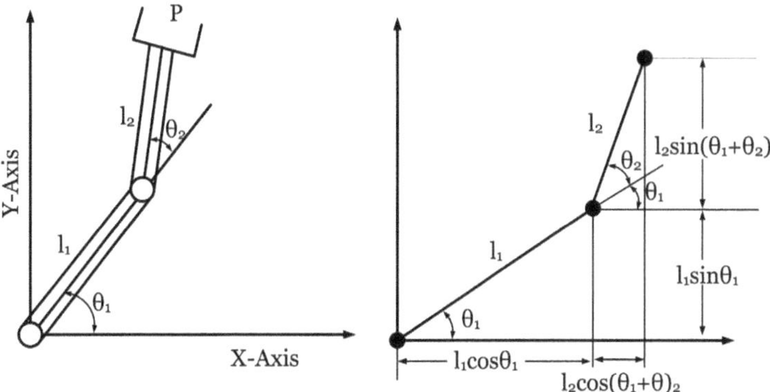

Fig. 6.6   2-DoF planer manipulator

From Eq. (6.12),

$$c\theta_2 = \frac{p_x^2 + p_y^2 - l_1^2 - l_2^2}{2l_1l_2} \tag{6.13}$$

Similarly

$$s\theta_2 = \pm\sqrt{1 - \left(\frac{p_x^2 + p_y^2 - l_1^2 - l_2^2}{2l_1l_2}\right)^2} \tag{6.14}$$

Finally, two possible solutions for θ_2 can be written as:

$$\theta_2 = A\tan 2\left(\pm\sqrt{1 - \left(\frac{p_x^2 + p_y^2 - l_1^2 - l_2^2}{2l_1l_2}\right)^2}, \frac{p_x^2 + p_y^2 - l_1^2 - l_2^2}{2l_1l_2}\right) \tag{6.15}$$

Multiplying each side of Eq. (6.8) by $c\theta_1$ and Eq. (6.9) by $s\theta_1$

$$c\theta_1 p_x = l_1 c^2\theta_1 + l_2 c\theta_1 c\theta_{12} \tag{6.16}$$

$$s\theta_1 p_y = l_1 s^2\theta_1 + l_2 s\theta_1\, s\theta_{12} \tag{6.17}$$

Adding the Eqs. (6.16) and (6.17)

$$c\theta_1 p_x + s\theta_1 p_y = l_1\left(c^2\theta_1 + s^2\theta_1\right) + l_2(c\theta_1 c\theta_{12} + s\theta_1 s\theta_{12}) \tag{6.18}$$

On simplification,

$$c\theta_1 p_x + s\theta_1 p_y = l_1 + l_2 c\theta_2 \tag{6.19}$$

Multiply each side of Eq. (6.8) by $-s\theta 1$ and Eq. (6.9) by $c\theta 1$ and add the resulting equations:

$$-s\theta_1\, p_x = -l_1\, s\theta_1.c\theta_1 - l_2\, s\theta_1 c\theta_{12}$$

$$c\theta_1\, p_y = l_1\, c\theta_1 s\theta_1 + l_2\, c\theta_1 s\theta_{12}$$

$$-s\theta_1\, p_x + c\theta_1\, p_y = l_2\left(c\theta_1\, s\theta_{12} - s\theta_1 c\theta_{12}\right) = -l_2\, s\theta_2\left(c^2\theta_1 + s^2\theta_1\right)$$

$$-s\theta_1\, p_x + c\theta_1\, p_y = l_2 s\theta_2 \tag{6.20}$$

Now, multiply each side of Eq. (6.19) by p_x and Eq. (6.20) by p_y and add the resulting equations:

$$c\theta_1 \left(p_x^2 + p_y^2 \right) = p_x \left(l_1 + l_2\, c\theta_2 \right) + p_y\, l_2\, s\theta_2 \tag{6.21}$$

$$c\theta_1 = \frac{p_x(l_1 + l_2 c\theta_2) + p_y l_2 s\theta_2}{p_x^2 + p_y^2} \tag{6.22}$$

and

$$s\theta 1 = \pm \sqrt{1 - \left(\frac{p_x(l_1 + l_2 c\theta_2) + p_y l_2 s\theta_2}{p_x^2 + p_y^2} \right)^2} \tag{6.23}$$

The possible result for θ_1 can be written as:

$$\theta_1 = A\tan 2\left(\pm \sqrt{1 - \left(\frac{p_x(l_1 + l_2 c\theta_2) + p_y l_2 s\theta_2}{p_x^2 + p_y^2} \right)^2} ,\ \frac{p_x(l_1 + l_2 c\theta_2) + p_y l_2 s\theta_2}{p_x^2 + p_y^2} \right) \tag{6.24}$$

This is how one can perform inverse kinematics approach for a 2-link planar arm in a geometric solution approach. But the difficulty level will increase with the increase in the number of links. Therefore, algebraic approach is preferred to perform inverse kinematics for manipulator with more than three links.

6.3.2 Algebraic Solution Approach

Application of this approach to a 6-DoF manipulator with revolute joints q_i and link lengths l_i ($i = 1, 2, 3, 4, 5, 6$) is considered for discussion. The homogeneous transformation matrix (HTM) for the manipulator can be written as:

$$T = \begin{bmatrix} r_{11} & r_{12} & r_{13} & p_x \\ r_{21} & r_{22} & r_{23} & p_y \\ r_{31} & r_{32} & r_{33} & p_z \\ 0 & 0 & 0 & 1 \end{bmatrix}$$

The closure equation of the manipulator is

$$T = T_1.T_2.T_3.T_4.T_5.T_6$$

The above equation is rewritten as:

$$^0_6T = {}^0_1T(q_1)\, {}^1_2T(q_2)\, {}^2_3T(q_3)\, {}^3_4T(q_4)\, {}^4_5T(q_5)\, {}^5_6T(q_6) \tag{6.25}$$

To find the inverse kinematics solution for the first joint (q_1) as a function of the known elements of $^{base}_{end-effector}T$, the link transformation inverse is premultiplied as follows:

$$\left[\left[{}^0_1T(q^0_1T(q_1))\right] - 1\right] {}^0_6T = \left[\left[{}^0_1T(q_1)\right] - 1\right] {}^0_1T(q_1)\, {}^1_2T(q_2)\, {}^2_3T(q_3)\, {}^3_4T(q_4)\, {}^4_5T(q_5)\, {}^5_6T(q_6)$$

With $[{}^0_1T(q1)]-1\, {}^0_1T(q1) = I$, I being the identity matrix, the above equation becomes:

$$\left[{}^0_1T(q1)\right] - 1\, {}^0_6T = {}^1_2T(q_2)\, {}^2_3T(q_3)\, {}^3_4T(q_4)\, {}^4_5T(q_5)\, {}^5_6T(q_6) \tag{6.26}$$

The following equations can be obtained in a similar manner.

$$\left[{}^0_1T(q_1)\, {}^1_2T(q_2)\right] - 1\, {}^0_6T = {}^2_3T(q_3)\, {}^3_4T(q_4)\, {}^4_5T(q_5)\, {}^5_6T(q_6) \tag{6.27}$$

$$\left[{}^0_1T(q_1)\, {}^1_2T(q_2)\, {}^2_3T(q_3)\right] - 1\, {}^0_6T = {}^3_4T(q_4)\, {}^4_5T(q_5)\, {}^5_6T(q_6) \tag{6.28}$$

$$\left[{}^0_1T(q_1)\, {}^1_2T(q_2)\, {}^2_3T(q_3)\, {}^3_4T(q_4)\right] - 1\, {}^0_6T = {}^4_5T(q_5)\, {}^5_6T(q_6) \tag{6.29}$$

$$\left[{}^0_1T(q_1)\, {}^1_2T(q_2)\, {}^2_3T(q_3)\, {}^3_4T(q_4)\, {}^4_5T(q_5)\right] - 1{}^0_6T = {}^5_6T(q_6) \tag{6.30}$$

There are 12 simultaneous set of nonlinear equations to be solved. The only unknown on the left-hand side of equation is q_1. The 12 nonlinear matrix elements of the right-hand side are zero, constant, or functions of q_2 through q_6. If the elements on the left-hand side which are the functions of q_1 are equated with the elements on the right-hand side, then the joint variable q_1 can be estimated as functions of r_{11}, $r_{12}, \ldots, r_{33}$, p_x, p_y, p_z and the fixed linked parameters.

On estimation of q_1, the other joint variables can be solved following the same method. Some trigonometric equations used in the solution of inverse kinematics problem are tabulated in Table 6.2.

Table 6.2 Some solution of inverse kinematics problem

Equations	Solutions
$a\ \sin\theta + b\ \cos\theta = c$	$\theta = A\ \tan 2(a, b) \mp A\ \tan 2\left(\sqrt{a^2 + b^2 - c^2}, c\right)$
$a\sin\theta + b\ \cos\theta = 0$	$\theta = A\ \tan 2(-b, a)$ or $\theta = A\ \tan 2(b, -a)$
$\cos\theta = a$ *and* $\sin\theta = b$	$\theta = A\ \tan 2(b, a)$
$\cos\theta = a$	$\theta = A\ \tan 2\left(\mp\sqrt{1 - a^2}, a\right)$
$\sin\theta = a$	$\theta = A\ \tan 2\left(a, \mp\sqrt{1 - a^2}\right)$

6.4   Jacobian Matrix

Jacobian matrix, used in the kinematic and dynamic analysis of a robot, relates the relation of joint velocities to the linear or angular velocities of end-effector. For the serial manipulator in Fig. 6.3, velocity of the end-effector using Jacobian can be expressed as:

$$\text{Velocity, V} = \begin{bmatrix} \dot{p}_x \\ \dot{p}_y \\ \dot{p}_z \end{bmatrix} = \begin{bmatrix} v \\ \omega \end{bmatrix} = J \begin{bmatrix} \dot{\theta}_1 \\ \dot{\theta}_2 \\ \cdot \\ \cdot \\ \cdot \end{bmatrix}$$

where,

$p_x,\ p_y,\ p_z$ are the position vectors
v is the joint linear velocity
ω is the joint angular velocity
$\theta_1,\ \theta_2\ \ldots$ are the joint variables (joint angle for revolute joints, displacement for prismatic joints)

6.4.1   Calculation of Jacobian

There are two different approaches to calculate Jacobian:

 (i) Partial differentiation method
(ii) Velocity propagation method

6.4.1.1   Partial Differentiation Method

Let us consider the following position equation for a robotic system.

$$y_1 = f_1 \left(x_1, x_2, x_3, \ldots\ldots\ldots\ldots\ldots\ldots\ldots\ldots\ldots\ldots \right) \qquad (6.31a)$$

$$y_2 = f_2 \left(x_1, x_2, x_3, \ldots\ldots\ldots\ldots\ldots\ldots\ldots\ldots\ldots\ldots \right) \qquad (6.31b)$$

$$y_n = f_n \left(x_1, x_2, x_3, \ldots\ldots\ldots\ldots\ldots\ldots\ldots\ldots\ldots\ldots \right) \qquad (6.31c)$$

Differentiating above equations, we get

$$\delta y_1 = \frac{\delta f_1}{\delta x_1} x_1 + \frac{\delta f_1}{\delta x_1} \delta x_2 + \ldots\ldots\ldots\ldots\ldots\ldots\ldots\ldots\ldots\ldots \qquad (6.32a)$$

$$\delta y_2 = \frac{\delta f_2}{\delta x_1} x_1 + \frac{\delta f_2}{\delta x_1} \delta x_2 + \ldots\ldots\ldots\ldots\ldots\ldots\ldots\ldots\ldots\ldots \qquad (6.32b)$$

$$\delta y_n = \frac{\delta f_n}{\delta x_1} x_1 + \frac{\delta f_n}{\delta x_1} \delta x_2 + \ldots\ldots\ldots\ldots\ldots\ldots\ldots\ldots\ldots\ldots \qquad (6.32c)$$

These sets of equations can be written in the form:

$$\begin{bmatrix} \dot{y}_1 \\ \dot{y}_2 \\ \cdot \\ \cdot \end{bmatrix} = \begin{bmatrix} & J & \end{bmatrix} \begin{bmatrix} \dot{x}_1 \\ \dot{x}_2 \\ \cdot \\ \cdot \end{bmatrix} \qquad (6.33)$$

which can be expressed as

$$\dot{Y} = J \dot{X} \qquad (6.34)$$

The matrix J is called the Jacobian of the system. The following example is considered for understand:

Let, position vector be, $\mathbf{P} = \begin{bmatrix} r_1 \cos\theta \\ r_2 \sin\theta \\ 0 \end{bmatrix}$

$$\dot{p}_x = -r_1 \sin\theta . \dot{\theta}$$

$$\dot{p}_y = r_2 \cos\theta . \dot{\theta}$$

$$\dot{p}_z = 0$$

Writing these equations in matrix form:

$$\begin{bmatrix} \dot{p}_x \\ \dot{p}_y \\ \dot{p}_z \end{bmatrix} = \begin{bmatrix} -r_1 \sin\theta \\ r_2 \cos\theta \\ 0 \end{bmatrix} \begin{bmatrix} & \dot{\theta} & \end{bmatrix}$$

Comparing this with the Eq. (6.34):

Fig. 6.7 A planar
manipulator

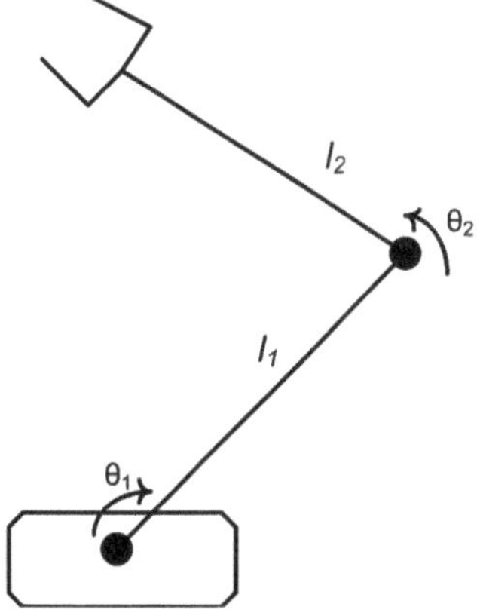

$$J = \begin{bmatrix} -r_1 sin\theta \\ r_2 cos\theta \\ 0 \end{bmatrix}$$

One can implement the following steps to find the Jacobian:

Step 1: Find the DH parameter of each link.
Step 2: Depending on the DH parameter, find the transformation matrix for each link.
   Thereafter, find the transformation matrix of end-effector with respect to the base.
Step 3: Use Eq. (6.34) to find the Jacobian (J).

Example 6.2 Find the velocity of the end-effector of the manipulator given in
Fig. 6.7.

 (i) Give the answer in terms of frame 0.
(ii) Give the answer in terms of frame 2.

Solution (i) DH parameter of the manipulator in Fig. 6.7

Link	b_i	θ_i	a_i	α_i
1	0	θ_1	l_1	0
2	0	θ_2	l_2	0

The transformation matrices are:

$$\begin{aligned}
{}^{0}_{1}T = \begin{bmatrix} c_1 & -s_1 & 0 & l_1 \\ s_1 & c_1 & 0 & 0 \\ 0 & 0 & 1 & 0 \\ 0 & 0 & 0 & 1 \end{bmatrix} \qquad
{}^{1}_{2}T = \begin{bmatrix} c_2 & -s_2 & 0 & l_2 \\ s_2 & c_2 & 0 & 0 \\ 0 & 0 & 1 & 0 \\ 0 & 0 & 0 & 1 \end{bmatrix}
\end{aligned}$$

$$
{}^{0}_{2}T = {}^{0}_{1}T \cdot {}^{1}_{2}T = \begin{bmatrix} c_{12} & -s_{12} & 0 & l_2 c_2 + l_1 \\ s_{12} & c_{12} & 0 & l_2 s_1 \\ 0 & 0 & 1 & 0 \\ 0 & 0 & 0 & 1 \end{bmatrix} \tag{6.35}
$$

The position vector of the end-effector can be achieved as:

$$
\mathbf{P} = \begin{bmatrix} l_2 c_2 + l_1 \\ l_2 s_1 \\ 0 \end{bmatrix}
$$

And,

$$\dot{p}_x = -l_2 s_1 \dot{\theta}_1$$

$$\dot{p}_y = l_2 c_1 \dot{\theta}_1$$

$$\dot{p}_z = 0$$

Writing these equations in terms of a matrix:

$$
\begin{bmatrix} \dot{p}_x \\ \dot{p}_y \\ \dot{p}_z \end{bmatrix} = \begin{bmatrix} -l_2 s_1 \\ l_2 c_1 \\ 0 \end{bmatrix} \begin{bmatrix} \dot{\theta}_1 \end{bmatrix} \tag{6.36}
$$

Comparing this Eq. (6.36) with Eq. (6.34),

$$
\text{Velocity vector}, \mathbf{V} = \begin{bmatrix} -l_2 s_1 \\ l_2 c_1 \\ 0 \end{bmatrix} \begin{bmatrix} \dot{\theta}_1 \end{bmatrix}
$$

$$
\text{Jacobian}, \mathbf{J} = \begin{bmatrix} -l_2 s_1 \\ l_2 c_1 \\ 0 \end{bmatrix} \tag{6.37}
$$

(ii) Further, to find velocity with respect to frame 2,

${}^{2}V = {}^{2}J \left[\dot{\theta} \right]$, where ${}^{2}J$ is Jacobian with respect to frame 2.

${}^{2}J = {}^{2}_{0}R \cdot {}^{0}J$, where ${}^{2}_{0}R$ is rotational matrix and ${}^{0}J$ is Jacobian with respect to frame 0.

Now,

$$
{}^{2}_{0}R = {}^{0}_{2}R^{-1} = \begin{bmatrix} c_{12} & s_{12} & 0 \\ -s_{12} & c_{12} & 0 \\ 0 & 0 & 1 \end{bmatrix}
\tag{6.38}
$$

Therefore,

$$
{}^{2}J = {}^{2}_{0}R.\ {}^{0}J = \begin{bmatrix} c_{12} & s_{12} & 0 \\ -s_{12} & c_{12} & 0 \\ 0 & 0 & 1 \end{bmatrix}\begin{bmatrix} -l_{2}s_{1} \\ l_{2}c_{1} \\ 0 \end{bmatrix} = \begin{bmatrix} -l_{2}c_{12}s_{1} + l_{2}s_{12}c_{1} \\ l_{2}s_{12}s_{1} + l_{2}c_{12}c_{1} \\ 0 \end{bmatrix}
\tag{6.39}
$$

The velocity with respect to frame 2,

$$
{}^{2}V = {}^{2}J\left[\dot{\theta}\right] = \begin{bmatrix} -l_{2}c_{12}s_{1} + l_{2}s_{12}c_{1} \\ l_{2}s_{12}s_{1} + l_{2}c_{12}c_{1} \\ 0 \end{bmatrix}\left[\dot{\theta}\right]
\tag{6.40}
$$

6.4.1.2   Velocity Propagation Method

In this method, the velocities are calculated based on the previous link's velocity and thereby the Jacobian can be calculated. For example, let us consider the 3 link manipulator shown in Fig. 6.8.

To obtain the velocity of end-effector, one needs to find the velocity of

(i) Base frame (0th frame)
(ii) 1st frame

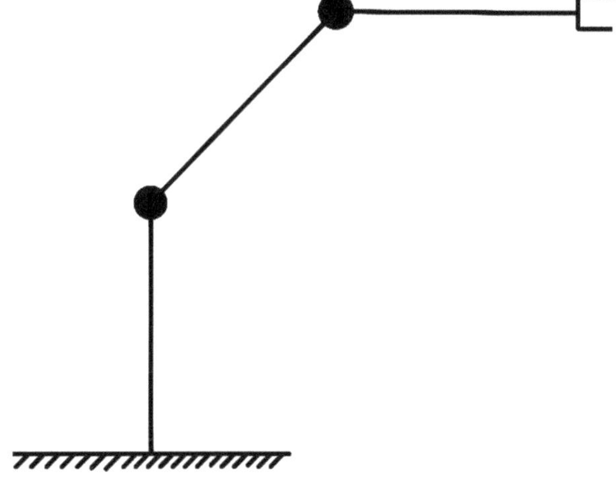

Fig. 6.8 A 3-link manipulator

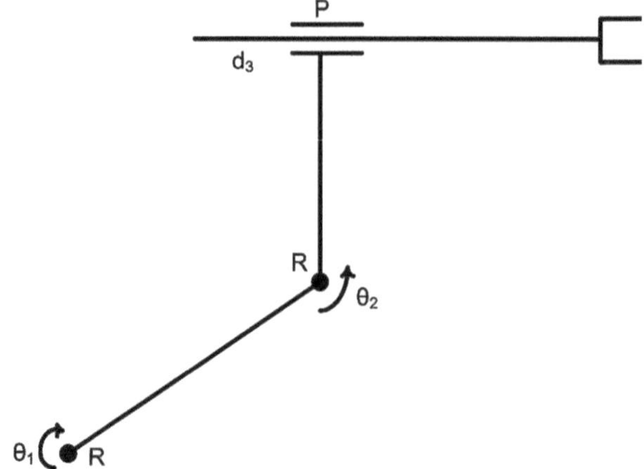

Fig. 6.9 A robotic system

(iii) 2nd frame
(iv) 3rd frame

The equations for finding the velocities are:

$$^{i+1}\omega_{i+1} = {}^{i+1}_{i}R.{}^{i}\omega_i + \dot{\theta}_{i+1}.{}^{i+1}\widehat{Z}_{i+1} \tag{6.41}$$

$$^{i+1}V_{i+1} = {}^{i+1}_{i}R.\left({}^{i}V_i + {}^{i}\omega_i \times {}^{i}P_{i+1}\right) + \dot{d}_{i+1}.{}^{i+1}\widehat{Z}_{i+1} \tag{6.42}$$

where,

ω = angular velocity
V = linear velocity
$^{i}P_{i+1}$ = position Vector

$$\dot{\theta}_{i+1}.{}^{i+1}\widehat{Z}_{I+1} = \begin{bmatrix} 0 \\ 0 \\ \dot{\theta}_{i+1} \end{bmatrix} \text{ (for revolute joint)}$$

$$\dot{d}_{i+1}.{}^{i+1}\widehat{Z}_{I+1} = \begin{bmatrix} 0 \\ 0 \\ \dot{d}_{i+1} \end{bmatrix} \text{ (for prismatic joint)}$$

Example 6.3 Consider the manipulator in Fig. 6.9 and solve the questions

(a) Find the linear and angular velocity at end-effector frame
(b) Find the Jacobian at the end-effector frame
(c) Find the Jacobian at the base frame

Solution (i) We have

$$
{}^0_1T = \begin{bmatrix} c_1 & -s_1 & 0 & 0 \\ s_1 & c_1 & 0 & 0 \\ 0 & 0 & 1 & 0 \\ 0 & 0 & 0 & 1 \end{bmatrix} \qquad
{}^1_2T = \begin{bmatrix} -s_2 & c_2 & 0 & 0 \\ 0 & 0 & 1 & 0 \\ c_2 & -s_2 & 0 & 0 \\ 0 & 0 & 0 & 1 \end{bmatrix} \qquad {}^1_2T
$$

$$
= \begin{bmatrix} 1 & 0 & 0 & 0 \\ 0 & 0 & -1 & d_3 \\ 0 & 1 & 0 & 0 \\ 0 & 0 & 0 & 1 \end{bmatrix}
$$

Now,

$$
{}^0_3T = {}^0_1T\cdot{}^1_2T\cdot{}^2_3T = \begin{bmatrix} -c_1s_2 & s_1 & c_1c_2 & c_1c_2d_3 \\ -s_1s_2 & c_1 & s_1c_2 & s_1c_2d_3 \\ c_2 & 0 & s_2 & s_3d_3 \\ 0 & 0 & 0 & 1 \end{bmatrix} \tag{6.43}
$$

Since base frame is fixed,

$$
{}^0V_0 = {}^0\omega_0 = \begin{bmatrix} 0 \\ 0 \\ 0 \end{bmatrix} \tag{6.44}
$$

For i = 0,

$$
{}^1\omega_1 = {}^1_0R\cdot{}^0\omega_0 + \dot{\theta}_1\cdot\widehat{Z} = 0 + \dot{\theta}_1\cdot\widehat{Z} = \begin{bmatrix} 0 \\ 0 \\ \dot{\theta}_1 \end{bmatrix} \tag{6.45}
$$

$$
{}^1V_1 = {}^1_0R\left({}^0V_0 + {}^0\omega_0 \times {}^0P_1\right) + 0 = \begin{bmatrix} 0 \\ 0 \\ 0 \end{bmatrix} \tag{6.46}
$$

Now for i = 1

$$
\begin{aligned}
{}^2\omega_2 &= {}^2_1R\left({}^1\omega_1\right) + \dot{\theta}_2\widehat{Z} \\
&= {}^1_2R^{-1}\left({}^1\omega_2\right) + \dot{\theta}_2\widehat{Z} \\
&= \begin{bmatrix} -S_2 & 0 & C_2 \\ -C_2 & 0 & -S_2 \\ 0 & -1 & 0 \end{bmatrix} \begin{bmatrix} 0 \\ 0 \\ \dot{\theta}_1 \end{bmatrix} + \begin{bmatrix} 0 \\ 0 \\ \dot{\theta}_2 \end{bmatrix} = \begin{bmatrix} C_2\dot{\theta}_1 \\ -S_2\dot{\theta}_1 \\ \dot{\theta}_2 \end{bmatrix}
\end{aligned} \tag{6.47}
$$

$$^2V_2 = {}^2_1R\left({}^1V_1 + {}^1\omega_1 \times {}^1P_2\right) + 0$$

$$= \begin{bmatrix} -S_2 & 0 & C_2 \\ -C_2 & 0 & -S_2 \\ 0 & -1 & 0 \end{bmatrix} \left(\begin{bmatrix} 0 \\ 0 \\ 0 \end{bmatrix} + \begin{bmatrix} 0 \\ 0 \\ \dot\theta_1 \end{bmatrix} \times \begin{bmatrix} 0 \\ 0 \\ 0 \end{bmatrix} \right) = \begin{bmatrix} 0 \\ 0 \\ 0 \end{bmatrix} \tag{6.48}$$

Now i = 2,

$$^3\omega_3 = {}^3_2R\left({}^2\omega_2\right) + 0$$

$$= \begin{bmatrix} 1 & 0 & 0 \\ 0 & 0 & 1 \\ 0 & -1 & 0 \end{bmatrix} \begin{bmatrix} C_2\dot\theta_1 \\ -S_2\dot\theta_1 \\ \dot\theta_2 \end{bmatrix} = \begin{bmatrix} C_2\dot\theta_1 \\ \dot\theta_2 \\ -S_2\dot\theta_1 \end{bmatrix} \tag{6.49}$$

$$^3V_3 = {}^3_2R\left[{}^2V_2 + {}^2\omega_2 \times {}^2P_2\right] + d_3\widehat{Z}_3$$

$$= \begin{bmatrix} 1 & 0 & 0 \\ 0 & 0 & 1 \\ 0 & -1 & 0 \end{bmatrix} \left(\begin{bmatrix} 0 \\ 0 \\ 0 \end{bmatrix} + \begin{bmatrix} C_2\dot\theta_1 \\ -S_2\dot\theta_1 \\ \dot\theta_2 \end{bmatrix} \times \begin{bmatrix} 0 \\ -d_3 \\ 0 \end{bmatrix} \right) + \begin{bmatrix} 0 \\ 0 \\ \dot d_3 \end{bmatrix} \tag{6.50}$$

$$= \begin{bmatrix} d_3\dot\theta_2 \\ -d_3C_2\dot\theta_1 \\ \dot d_3 \end{bmatrix}$$

The linear velocity at end-effector frame, $^3V_3 = \begin{bmatrix} d_3\dot\theta_2 \\ -d_3C_2\dot\theta_1 \\ \dot d_3 \end{bmatrix}$

The angular velocity at end-effector frame, $^3\omega_3 = \begin{bmatrix} C_2\dot\theta_1 \\ \dot\theta_2 \\ -S_2\dot\theta_1 \end{bmatrix}$

(b) To find the Jacobian in the final frame:

$$^3V = \begin{bmatrix} d_3\dot\theta_2 \\ -d_3C_2\dot\theta_1 \\ \dot d_3 \end{bmatrix} = \begin{bmatrix} 0 & d_3 & 0 \\ -d_3C_2 & 0 & 1 \\ 0 & -1 & 0 \end{bmatrix} \begin{bmatrix} \dot\theta_1 \\ \dot\theta_2 \\ \dot d_3 \end{bmatrix} \tag{6.51}$$

Comparing it with Y=J$\dot x$, we get,

$$^3J = \begin{bmatrix} 0 & d_3 & 0 \\ -d_3C_2 & 0 & 1 \\ 0 & -1 & 0 \end{bmatrix}$$

This is the Jacobian matrix at the frame at the end-effector frame.

(c) To find the Jacobian at the base frame

$$
{}^{0}J = {}^{0}_{3}R\,{}^{3}J
$$

$$
=
\begin{bmatrix}
-C_1 S_2 & S_1 & C_1 C_2 \\
-S_2 & C_1 & S_1 C_2 \\
C_2 & 0 & S_2
\end{bmatrix}
\begin{bmatrix}
0 & d_S & 0 \\
-d_3 C_2 & 0 & 0 \\
0 & 0 & 1
\end{bmatrix}
$$

$$
=
\begin{bmatrix}
-S_1 d_3 C_2 & -C_1 S_2 d_3 & C_1 C_2 \\
-C_1 d_3 C_2 & -S_1 S_2 d_3 & S_1 C_2 \\
0 & C_2 d_3 & S_2
\end{bmatrix}
\tag{6.52}
$$

This is the Jacobian matrix at frame 0 or base frame.

6.5   Forward and Inverse Dynamics

Dynamics is the study of forces or moments causing the motion in a system. Mathematical models of a robot's dynamics provide a description of why things move when forces are generated in or applied on the system. This dynamical behavior of a robot can be expressed by a set of equations called equations of motion, which govern the dynamic response of the robot linkage with respect to the input joint torques. Dynamic models can be used to develop suitable control strategies and simulation of a robotic system. The dynamic analysis of a robot gives all the joint reaction forces and moments needed for the design and sizing of the links, bearings, and actuators.

Let ($q, \dot{q}$) be the dynamic state and $u \in R_n$ the controls (typically joint torques in each motor) of a robot. Robot dynamics can be written as:

$$
M(q)\ddot{q} + C(q, \dot{q})\,\dot{q} + G(q) = u
\tag{6.53}
$$

$M(q) \in R_n \times n$ is positive definite intertia matrix (can be inverted into forward simulation of dynamics). The inertia term depends on the mass distribution of the robotic link and is expressed in terms of moment of inertia of that matrix.

$C(q, \dot{q}) \in R_n$ are the centripetal and the coriolis forces. The coriolis components will appear whenever there is a sliding joint on a rotary link.

$G(q) \in R^n$ are the gravitational forces

u is the joint torque

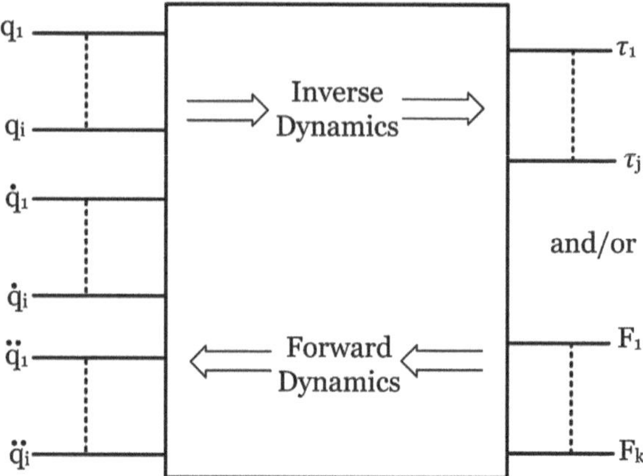

Fig. 6.10 Schematic of forward and inverse dynamics

We often write more compactly as:

$$M(q)\ddot{q} + F(q, \dot{q}) = u \tag{6.54}$$

In case of a robot, knowing its physical parameters, one normally wishes to solve two problems related to its dynamics. They are forward and inverse dynamics. The schematic of forward and inverse dynamics is presented in Fig. 6.10.

Forward dynamics computes joint motions of a robot for a given set of joint torques or forces as a function of time. Forward dynamics is required to find the response of the robot arm corresponding to the applied torques or forces at the joints. It is used primarily for computer simulation of a robot, which just shows how a robot will perform when it is built.

In case of forward dynamics, if the applied torques are known, one can use the equation below to simulate the dynamics of the system:

$$\ddot{q} = M(q)^{-1}(u - F(q, \dot{q})) \tag{6.55}$$

Inverse dynamics deals with the evaluation of joint torques and forces required for a set of joint motions for a particular force at the end-effector. The inverse dynamics problem is used to find the actuator torques or forces required to generate a desired trajectory of the robot's end-effector. An efficient inverse dynamics model becomes extremely important for real-time control of robots.

6.5.1 Dynamic Equation of Motion

Two basic approaches used for dynamic analysis of a robot are considered for discussion in this section: Euler-Lagrange and Newton-Euler equations of motion.

Table 6.3 Euler-Lagrange method vis-à-vis Newton-Euler

Euler-Lagrange method	Newton-Euler method
Energy-based approach	Vector-based approach
Dynamic equations in closed form	Dynamic equations in numeric/recursive form
Often used for study of dynamic properties and analysis of control methods	Often used for numerical solution of forward/inverse dynamics

The Newton-Euler formulation is derived by the direct interpretation of Newton's second law of motion, which describes dynamic systems in terms of force and momentum. It incorporates all the forces and moments acting on the individual robot links, including the coupling forces and moments between the links. The equations obtained from the Newton-Euler method include the constraint forces acting between adjacent links. Thus, additional arithmetic operations are required to obtain explicit relations between the joint torques and the resultant motion in terms of joint displacements.

On the other hand, in the Euler-Lagrange method, the system's dynamic behavior is described in terms of work and energy using generalized coordinates. Therefore, all the workless forces and constraint forces are automatically eliminated. The resultant equations are generally compact and provide a closed-form expression in terms of joint torques and joint displacements. Furthermore, the derivation is simpler and more systematic than in the Newton-Euler method (Table 6.3).

6.5.1.1   Euler-Lagrange Formulation

Euler-Lagrange method describes the behavior of a dynamic system in terms of work and energy stored in the system. The constraint forces involved in the system are automatically eliminated in the formulation of Lagrangian dynamic equations. Using the concept of generalized coordinates and Lagrangian, the dynamic model of a robot can be derived in a systematic way. The Lagrangian (L) is defined as the difference between the kinetic and potential energies of a system.

$$L = T - U \tag{6.56}$$

where L is Lagrangian, T is the total kinetic energy, and U is total potential energy of the system.

The kinetic energy depends on both configurations, that is, (a) position and orientation and (b) velocity of the links of a robotic system, whereas the potential energy depends only on the configuration of the links. In such a case, the Euler-Lagrange equations of motion are given by

$$\frac{d}{dt}\left(\frac{\partial L}{\partial \dot{q}_i}\right) - \frac{\partial L}{\partial q_i} = \varnothing_i, \quad \text{for } i = 1, ..n \tag{6.57}$$

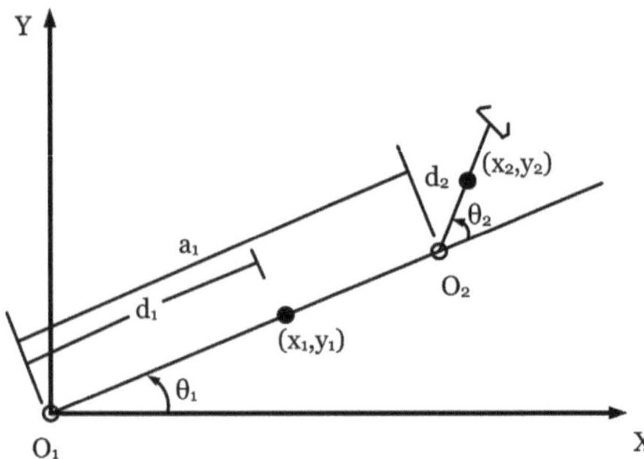

Fig. 6.11 A two-link robot arm

where n is the number of independent generalized coordinates used to define the system's configuration and q_i and $\varnothing_i$ are the generalized coordinates and generalized forces due to applied forces corresponding to the generalized coordinates, respectively.

Generalized Coordinates

The coordinates that specify the configuration, i.e., the position and orientation of all the bodies or links of a mechanical system completely, are called generalized coordinates. Since a rigid link on a plane has three DoFs, a mechanical system with m moving links requires $3m$ coordinates to specify its configuration completely in the plane.

A two-link planar robot as shown in Fig. 6.11 is considered for example. Since a rigid link on a plane has three DoF, two links require six coordinates, namely, (x_1, y_1, θ_1) and (x_2, y_2, θ_2) which are not independent as there are two revolute joints that restrict the motion of the two bodies.

The coordinates (x_1, y_1) and (x_2, y_2) define the positions of the center of mass of the links, whereas θ_1 and θ_2 denote the orientation of the links. Moreover, d_1 and d_1 are the mass center locations from the origin of the frames. For the first set of six coordinates, there are four constraints:

$$x_1 = d_1 \cos \theta_1; y_1 = d_1 \sin \theta_1 \tag{6.58}$$

$$x_2 = a_1 \cos \theta_1 + d_2 \cos \theta_{12}; y_2 = a_1 \sin \theta_1 + d_2 \sin \theta_{12} \tag{6.59}$$

where $\theta_{12} = \theta_1 + \theta_2$. Therefore, the system has 6-4 = 2 DoF, and the independent set of generalized coordinates are θ_1 and θ_2.

Kinetic Energy

Consider a robot consisting of n rigid links as shown in Fig. 6.12a.

The kinetic energy of a typical link i shown in Fig. 6.12b denoted by T_i is given by

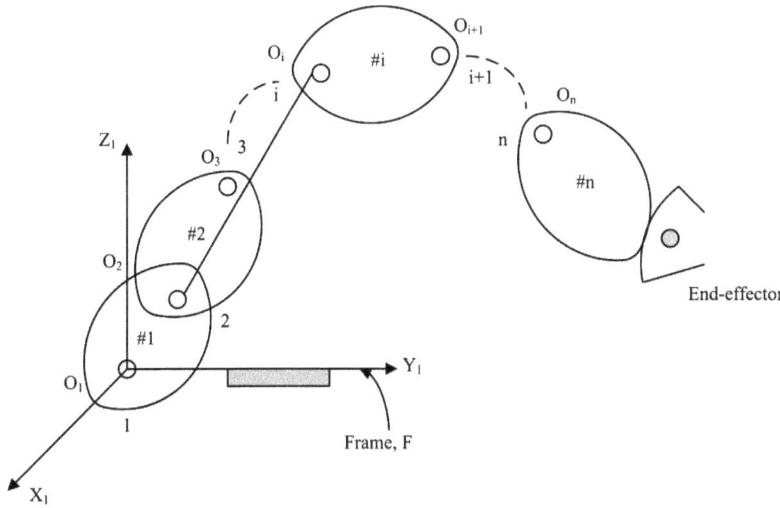

(a) A serial chain manipulator

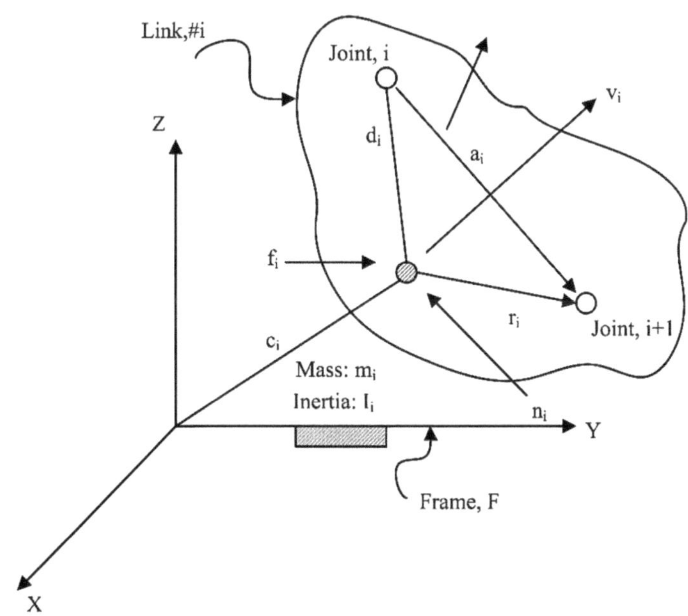

(b) The ith body (or link)

Fig. 6.12 A serial chain robot. (**a**) A serial chain manipulator, (**b**) The ith body (or link)

$$T_i = \frac{1}{2} m_i \dot{c}_i^T \dot{c}_i + \frac{1}{2} \omega_i^T I_i \omega_i \tag{6.60}$$

where

$\dot{c}_i = J_{c,i}\dot{\theta}$ is three-dimensional velocity vector of the mass center C_i of the ith link
$\omega_i = J_{\omega,i}\dot{\theta}$ is three-dimensional angular velocity vector of the ith link
m_i : Mass of the ith link (a scalar quantity)
I_i : The 3×3 inertia tensor or matrix of the ith link about C_i

The total kinetic energy of the robot is the sum of the contributions of each rigid link due to the relative motion and is given by

$$T = \frac{1}{2} \sum_{i=1}^{n} \left(m_i \dot{c}_i^T \dot{c}_i + \omega_i^T I_i \omega_i \right) = \frac{1}{2} \sum_{i=1}^{n} \dot{\theta}^T \overline{I}_i \, \dot{\theta} \tag{6.61}$$

where the $n \times n$ matrix $\overline{I}_i = m_i J_{c,i}^T J_{c,i} + J_{\omega,i}^T I_i J_{\omega,i}$ and $J_{c,i}^T J_{c,i}$ and $J_{\omega,i}^T I_i J_{\omega,i}$ are $n \times n$ matrices.

Moreover, if the $n \times n$ matrix is defined by

$$I = \sum_{i=1}^{n} \overline{I}_i \tag{6.62}$$

then the total kinetic energy can be rewritten as

$$T = \frac{1}{2} \dot{\theta}^T I \dot{\theta} \tag{6.63}$$

The matrix I is called the Generalized Inertia Matrix (GIM) of the robot.

Potential Energy
The potential energy stored in link i is defined as the amount of work required to raise the center of mass of link i from a horizontal reference plane to its present position under the influence of gravity. Similar to kinetic energy, the total potential energy stored in a robot is given by the sum of the contributions of each link and is given by

$$U = -\sum_{i=1}^{n} m_i c_i^T g \tag{6.64}$$

where g is the vector due to gravity acceleration and the vector c_i is a function of joint variables, i.e., θ_i's of the robot.

Equation of Motion
Using the values of potential and kinetic energy of the robot in Eq. (6.56), the Lagrangian is obtained as:

$$L = T - U = \sum_{i=1}^{n} \left[\frac{1}{2} \dot{\theta}^T \overline{I}_i \dot{\theta} + m_i c_i^T g \right] \tag{6.65}$$

Let i_{ij} be the (i, j) element of the robot's GIM I, then Eq. (6.65) can be written as

$$L = \sum_{i=1}^{n} \left[\sum_{j=1}^{n} \frac{1}{2} i_{ij} \dot{\theta}_i \dot{\theta}_j + m_i c_i^T g \right] \tag{6.66}$$

Next, the Lagrangian function is differentiated with respect to θ_i, $\dot{\theta}_i$, and t to obtain the dynamic equations of motion. After differentiating L with respect to θ_i, $\dot{\theta}_i$, and t, and then combining, the dynamic equation of motion is derived as:

$$\sum_{j=1}^{n} i_{ij} \theta_j + h_i + \gamma_i = \tau_i \tag{6.67}$$

for i = 1 where

$$h_i = \sum_{j=1}^{n} \sum_{k=1}^{n} \left(\frac{\partial i_{ij}}{\partial \theta_k} - \frac{1}{2} \frac{\partial i_{ij}}{\partial \theta_i} \right) \dot{\theta}_j \dot{\theta}_k \tag{6.68}$$

$$\gamma_i \equiv - \sum_{j=1}^{n} \left[j_{c,j}^{(i)} \right]^T (m_j g) \tag{6.69}$$

Writing Eq. (6.67) for all the n generalized coordinates, the equation of motion can be written in a compact form as:

$$I\ddot{\theta} + h + \gamma = \tau \tag{6.70}$$

Newton-Euler Formulation

The motion of a rigid body can be decomposed into translational motion with respect to an arbitrary point fixed to the rigid body, and the rotational motion of the rigid body about that point. The dynamic equations of a rigid body can also be represented by two equations: one describes the translational motion of the centroid (or center of mass), while the other describes the rotational motion about the centroid. The former is Newton's equation of motion for a mass particle, and the latter is called Euler's equation of motion.

As shown in Fig. 6.13, let F be the fixed frame. Moreover, the vector m is the linear momentum of the rigid link or the body B expressed in the frame F. The corresponding angular momentum is represented by the vector $\tilde{m}$. Also, let vectors f^o and n^o be the resultant forces and moments exerted on B at and about the origin O, respectively. Then, the Newton's linear equation of motion can be stated as the time-

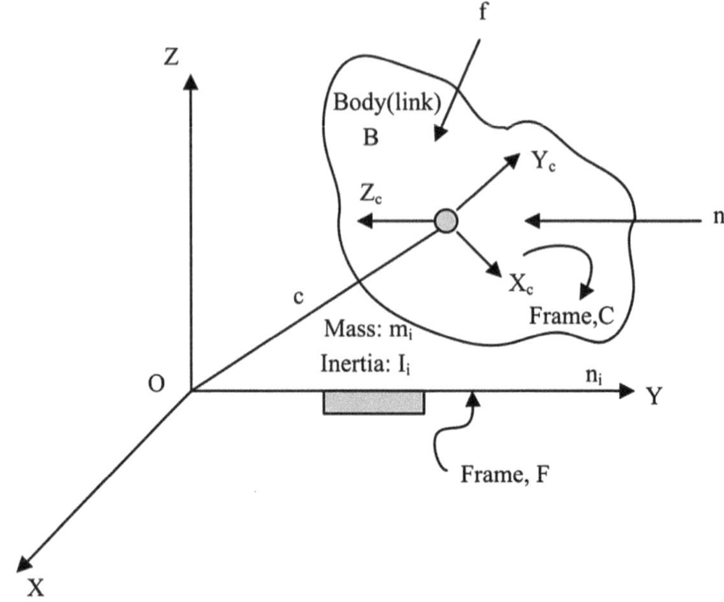

Fig. 6.13 Resulting force and moment acting on a rigid body (link)

derivative of the linear momentum m, which equals to the external forces acting on it, i.e.,

$$f^o = \frac{dm}{dt} \tag{6.71}$$

On the other hand, the Euler's equation of rotational motion gives the time rate of change of angular momentum $\tilde{m}^o$ to be equal to the external moment acting on it (remember the point about which the angular momentum and external moment are taken, which should be same), i.e.,

$$n^o = \frac{d\tilde{m}^o}{dt} \tag{6.72}$$

Again, linear momentum is given by

$$m = \frac{d}{dt}(mc) = m\dot{c} \tag{6.73}$$

For a body of constant mass substituting Eq. (6.73) into Eq. (6.71) gives

$$f^o = m\frac{d\dot{c}}{dt} = m\ddot{c} \tag{6.74}$$

Equation (6.74) is called Newton's equation of motion for the center of mass. Again, the angular momentum of a rigid body about its center of mass C is given by

$$\tilde{m} \equiv I\omega \tag{6.75}$$

where I is the inertia tensor of the body B about its center of mass C.

Substituting Eq. (6.75) into Eq. (6.72), we get

$$[n]_c = [I]_c[\dot{\omega}]_c + [\omega]_c \times \left([I]_c[\omega]_c\right) \tag{6.75}$$

Equation (6.75) is called Euler's equation of rotational motion for the center of mass coordinate frame.

For a robot to accomplish any task, the pose, i.e., position and orientation of the links, joints, and end–effector, is to be known. Kinematic analysis for determining the pose using the forward and inverse kinematic techniques presented in this chapter can be simulated using the software tools mentioned in Sect. 8.6.3 in Chap. 8. Further, these tools will also enable the reader to perform the dynamic analysis of a robot manipulator, which is required for their motion control.

Chapter 7
Control Systems in Robotics

A control system determines the behavior of a system to generate a desired response. This chapter describes the basic concepts of control systems in robotics, robot control techniques, and their domains of use.

7.1 Basic Concepts of a Control System

A control system changes the state of a system by managing, commanding, and regulating its behavior to produce a desired response. It consists of three components: input, logic operation, and output. The input is the command or signal which actuates the control system for performing a particular task. Logic operation is the strategic process to achieve a new output state based on the input. It is usually programmed in an embedded controller. The output parameter is the final response produced depending on the input and logic operation. Figure 7.1 shows a diagrammatic representation of a control system.

7.2 Classification of Control Systems

Natural and man-made control systems: The naturally present biological systems in a living being and the environment are natural control systems. The control systems developed by man are man-made control systems; e.g., the human digestive system is a natural control system, whereas an automobile engine is a man-made control system.

Combinational control systems: The fusion of natural control systems and man-made control systems forms combinational systems, e.g., a prosthetic hand control using body signals and electronic systems.

N. M. Kakoty et al., *Introduction to Embedded Systems and Robotics*,
https://doi.org/10.1007/978-3-031-73098-6_7

Fig. 7.1 Diagrammatic representation of a control system

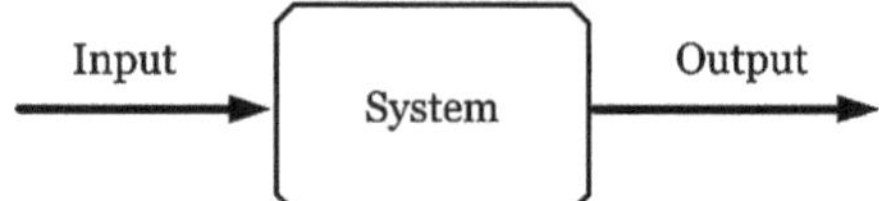

Time-varying and time-invariant control systems: Control systems with time-varying parameters are time-varying control systems, and control systems with parameters not varying with time are time-invariant control systems, e.g., the fuel unit of a space craft is a time-varying control system, whereas a weighing balance is a time-invariant control system. Fuel unit of a space craft is a time-varying control system because its parameters, such as fuel mass, propulsion requirements, and environmental conditions, change over time, necessitating real-time adjustments for effective control. In contrast, a weighing balance is a time-invariant control system as its parameters remain constant during operation, providing consistent measurements regardless of when they are taken.

Linear and non-linear systems: The control systems that satisfy the additive and homogeneous property are linear control systems whereas, those that do not satisfy the additive and homogeneous property are termed as non-linear systems; e.g., a refrigerator is a linear control system whereas, the speed control system in an automobile is an example of a non-linear control system. A refrigerator operates as a linear system because its cooling process is proportional to the control input, with a direct and predictable relationship. In contrast, speed control system of an automobile is non-linear due to factors like engine dynamics, aerodynamics, and varying road conditions, which create complex and non-proportional relationships between input and output.

Continuous-time and discrete-time control systems: Control systems having variation of output continuously as functions of input are called continuous control systems, and those control systems whose outputs vary at discrete intervals of time are called discrete time control systems, e.g., power supply regulator is a continuous time and stepper motor control system is discrete-time control system.

Deterministic and stochastic control systems: Control systems having a predictable output response are termed deterministic control systems and those with an unpredictable output response are termed stochastic control systems, e.g., a room heater is a deterministic and random number generator is a stochastic control system.

Single-input single-output (SISO) and multiple-input multiple-output (MIMO) control systems: Control systems having a single input and a single output are defined as SISO control systems, whereas control systems having multiple inputs and multiple outputs are defined as MIMO control systems, e.g., speedometer of an automobile is a SISO and air craft control is a MIMO system.

Open-loop and closed-loop systems: Based on the presence of a feedback system, a control system is divided into two types. Control systems having no feedback to the input from the output are open-loop control systems. The control action is therefore independent of the output, e.g., automatic washing machines and automatic coffee machine.

Control systems having feedback to the input from the output are closed-loop systems. The control action in this type is dependent on the output, e.g., temperature control system and missile launching system.

Feedback is an important tool in the control of any system. The presence of a feedback path provides an improved stability to the system.

7.3   Basic Concepts of Control System in Robotics

Some basic concepts for understanding a control system are state, estimate, reference, error, and dynamics.

7.3.1   State

State in a robotic control system refers to the output of the system. It depends on its previous states, the stimulus or input applied to the actuators of the robotic system, and the physics of the robot's environment. Pose, speed, velocity, angular velocity, and force are some of the states in a robotic control system.

7.3.2   Estimate

The exact state of a robotic system cannot be determined by the robot, but it can be estimated using sensors which equip the robot. Sensors with good accuracy, sensitivity, and precision are essential to produce a good estimation of the state.

7.3.3   Reference

Reference is the desired state to be reached in a robotic system.

7.3.4   Error

Error is the difference between the reference and the estimate state of a robotic system.

7.3.5   Dynamics

Dynamics in a robotic system describes the behavior of the system under non-static conditions with reference to time. Figure 7.2 shows a schematic of a control system. The state of the system is represented by x, the estimated state is represented by y, the control signal is represented by u, the reference is represented by r, and the error is represented by e. It is always the key responsibility of an engineer to build a controller that reacts and produces control signal **u,** such that **e~0** and **x~r.**

7.3.6   Types of Common Control Systems in Robotics

For developing a robotic system to perform a desired task, we require precise control of variables like pose, velocity, force, and torque. A robotic control system executes a planned sequence of motions and forces in the presence of unforeseen errors such as inaccuracies in the model of the robot, tolerances in the work piece, static friction in joints, mechanical compliance in linkages, electrical noise on transducer signals, and limitations in the precision of computation. Control systems allow every joint or wheel of the robot to follow a specific set of commanded motions and functions. Control systems in robotics are broadly categorized into two types as described in the following sections.

7.3.6.1   Open Control Loop

The systems in which the output has no effect on the control action are called open-loop control systems. In any open-loop system, the output isn't compared with the reference input. Thus, to every reference input, there corresponds a hard and fast operating condition. The accuracy of such systems depends on the calibration. Within the presence of disturbances, an open-loop system cannot perform the

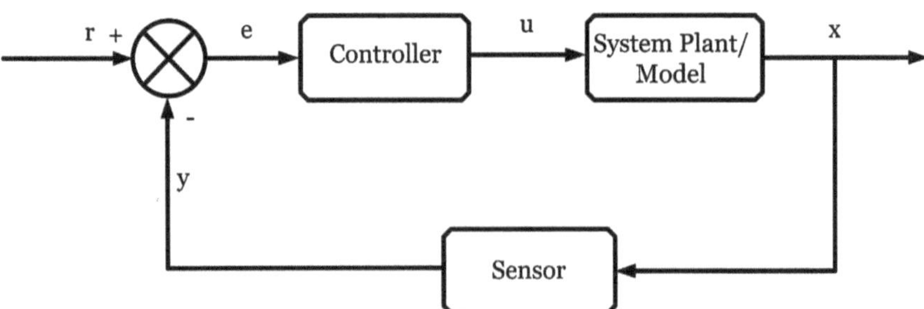

Fig. 7.2 Schematic of a robotic control system

required task because when the output changes due to disturbances, it is not followed by changes in its input to correct its output.

7.3.6.2 Feedback Control Loop

Feedback control systems are closed-loop control systems. The terms, *closed-loop control* and *feedback control,* are often used interchangeably. The actuating error signal which is the difference between the input and the feedback signal is fed to the controller to reduce the error and produce an output of the system to a desired value.

7.3.7 Basic Control Techniques in Robotics

7.3.7.1 Proportional Integral Derivative (PID) Controller

A PID controller is a closed-loop control system in robotics. The PID algorithm comprises of three coefficients—proportional, integral, and derivative. The controller reads a sensor to produce a desired actuator output by evaluating the proportional, integral, and derivative responses to the error. Its feedback mechanism helps in control of the output response. The application of PID controllers for automation is most common due to their accuracy, stability, flexibility, and reliability.

7.3.7.2 Linear Quadratic Controller (LQR)

Linear quadratic controller (LQR) is a procedure in present day control, which utilizes state space to deal with examination of a system. MIMO outline methodology is the ideal control strategy for LQR. The simple kind of loop shaping in scalar systems does not extend tomultivariable (MIMO) plants, which are characterized by transfer matrices rather than transfer functions. The notion of optimality is closely tied to MIMO system design. Optimal controllers, i.e., controllers that are consistent with some figures of merit end up to get only stabilizing controllers for MIMO plants. The linear quadratic regulator (LQR) is a well-known design technique that has practical feedback gains.

7.3.7.3 Predictive Controller

Predictive control is a control algorithm based on a predictive model of the process. The model is used to predict the future output based on historical information about the process, as well as anticipated future input. It emphasizes on the function of the model, and not the structure of the model.

7.3.7.4   Adaptive Control

Adaptive control uses feedback to update the model of the process based on the results of previous actions. The measurement of the results of previous actions is used to adapt the process model to correct for changes in the process. This type of adaption corrects for errors in the model due to long-term variations in the environment, but it cannot correct for dynamic changes caused by local disturbances.

7.3.8   Control Architectures in a Robotic System

Robot control architectures are conceptual structures for organizing robot control such that one can design controllers systematically. The term robot architecture is used to refer to how a system is divided into sub-systems and how the subsystems interact. Robot architectures and programming began in the late 1960 with the Shakey robot (schematic shown in Fig. 7.3) at Stanford University.

Shakey's architecture was decomposed into three functional elements: sensing, planning, and executing. The sensing system translated the camera image into an

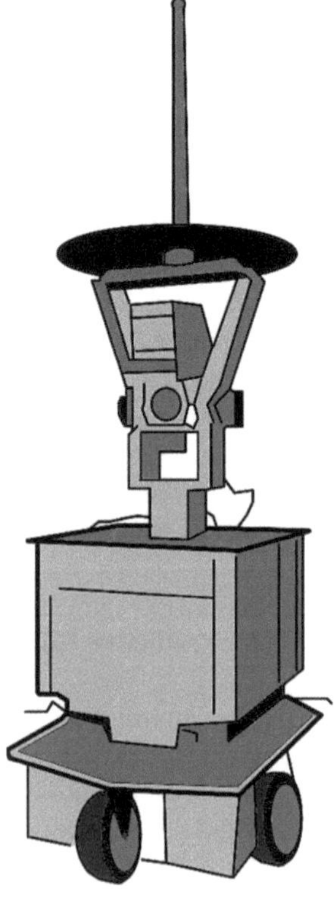

Fig. 7.3 Schematic of Shakey robot

internal world model. The planner took the internal world model and a goal and generated a plan (i.e., a series of actions). The executor took the plan and sent the actions to the robot's actuators.

7.3.8.1   Sense-Plan-Act Based Architecture

Figure 7.4 shows a schematic of the sense-plan-act based architecture for a robotic system. Sensing provides the robot with information about the state of itself and the environment. From this, a decision is made about how the robot should act and the controller commands the robot actuators to perform accordingly. The differences among robot control architectures lie almost entirely in the control logic, i.e., in the controller.

Robot control architecture can be divided into two classes: model-based architecture and sensor-based architecture.

7.3.8.2   Model-Based Architecture

Some robot control architectures use internal models to help the controller to decide what to do. These models are typically mathematical models, maps of the environment, or mechanical solid models. Model-based control architecture involves intensive processing which costs power and computational time. However, to store models used to represent robot control, architectures need memory. Further, model-based architectures need constant updates for applications in dynamic environment.

There are three common robot control approaches.

Hierarchical Approach
Hierarchical approach as presented in Fig. 7.5 makes extensive use of stored information and models to predict what might happen under different inputs, attempting to optimally choose an output. This allows the robot to plan a sequence of actions to achieve complex goals or exhibit a behavior, thereby allowing a

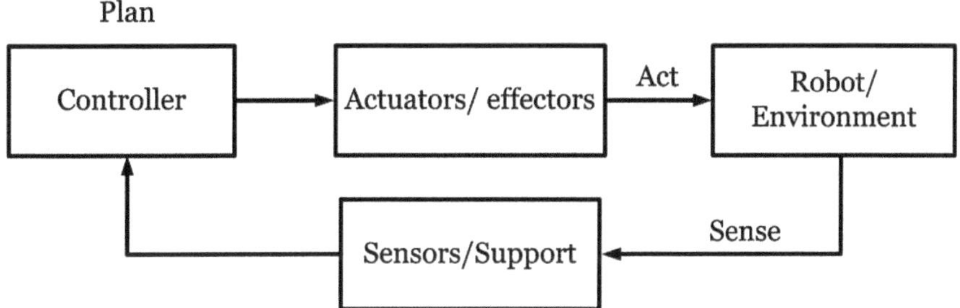

Fig. 7.4 Schematic of sense-plan-act structure common to all robot control architectures

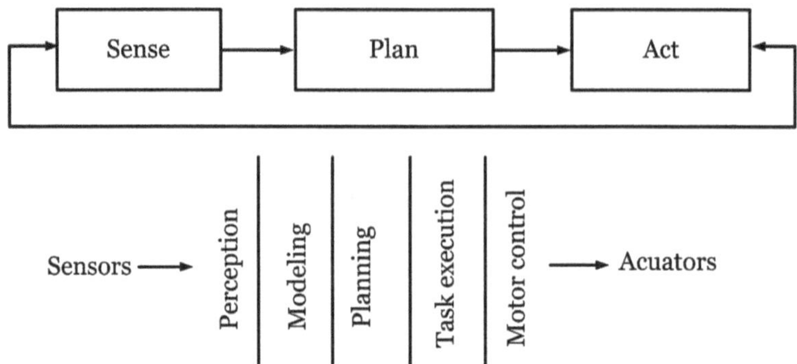

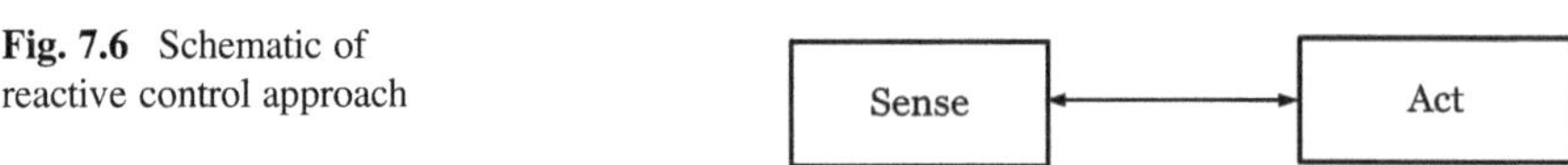

Fig. 7.5 Schematic of hierarchical control approach

Fig. 7.6 Schematic of
reactive control approach

designer to give commands that are interpreted in terms of the robot model. This paradigm is often called sense-plan-act (SPA), thereby substituting plan for decide in our usual scheme.

The control components in hierarchical control are said to be horizontally organized. Information from the environment in the form of sensor data has to filter through several intermediate stages of interpretation before finally becoming available for a response. The main architectural features of the SPA approach are that sensing flowed into a world model, which was then used by the planner, and that plan was executed without directly using the sensors that created the model.

The emphasis in these early systems was in constructing a detailed world model and then carefully planning out what steps to take next. The problem was that, while the robot was constructing its model and deliberating about what to do next, the world was likely to change. So, these robots exhibited the odd behavior that they would look (acquire data, often in the form of one or more camera images), process, and plan, and then (often after a considerable delay) they would result into action for a couple of steps before beginning the cycle all over again. Hierarchical architectures tend to support the evolution of intelligence from semi-autonomous control to fully autonomous control.

Reactive Approach

A reactive approach used in robotics control as presented in Fig. 7.6 is a connection of sensing with acting.

In 1986, Rodney A. Brooks published an article which described a type of reactive approach called the subsumption architecture. A subsumption architecture is built from layers of interacting finite-state machines, each connecting sensors to actuators directly. These finite-state machines were called behaviors (leading some

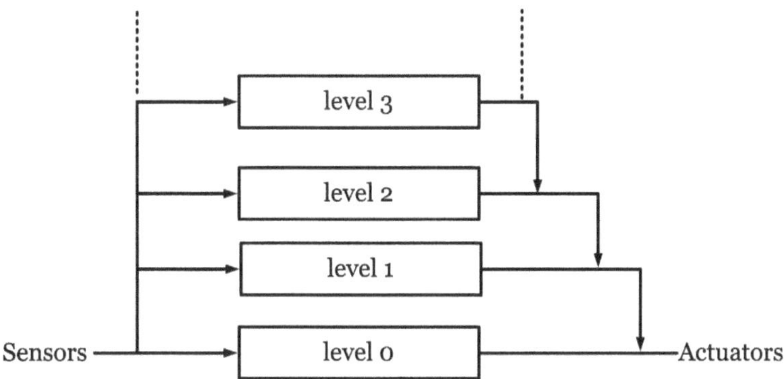

Fig. 7.7 Schematic of subsumption architecture

Fig. 7.8 Schematic of
hybrid control approach

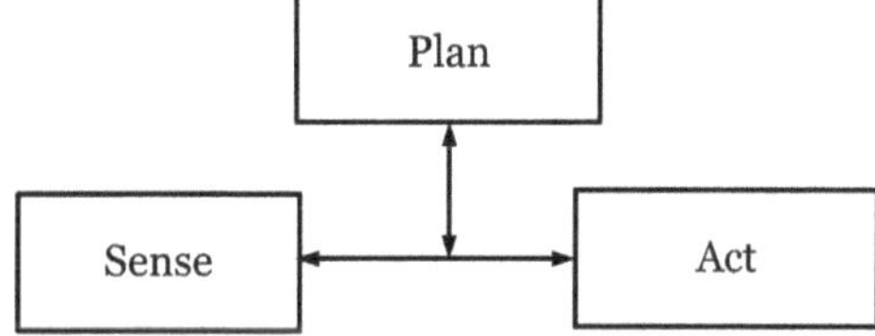

to call the subsumption architecture as behavior-based or behavioral robotics). Since multiple behaviors could be active at any one time, subsumption had an arbitration mechanism that enabled higher-level behaviors to override signals from lower-level behaviors. The subsumption architecture as in Fig. 7.7 became the dominant approach within the reactive robot control approaches.

The subsumption architecture by Brooks was characterized by:

1. A lack of representation of the outside world
2. The analysis of the architecture on a task rather than a functional basis
3. The subsuming of behaviors by higher level behaviors
4. A tight coupling of sensors and actuators

Hybrid Approach

Hybrid control approach shown in Fig. 7.8 combines the hierarchical and reactive control approach.

Hybrid approach can be described as plan, then sense, and act. Planning covers a long-time horizon and it uses global world model. Sense-act covers the reactive (real-time) part of the control. Hybrid approach may be characterized by a layering of capabilities, where low-level layers provide reactive capabilities and high-level layers provide the more computationally intensive deliberative capabilities.

The most popular variant on the hybrid approach is three layered: controller or reactive layer, sequencer or executive layer, and planner or deliberative layer. The controller or reactive layer provides low-level control of the robot. It is characterized by a sensor-action loop. Its decision cycle is often in the order of milliseconds.

7.3.8.3 Sensor-Based Architecture

Robotic sensors are used to estimate a robot's internal condition and surrounding environment. The environmental information is passed to a controller through sensors to enable appropriate behavior by the robot. The controller acts as the brain of the robot and has the instructions programmed by the roboticists. Figure 7.9 represents the types of sensors used in robotics.

Internal sensors are used to collect robot's information like joint position, velocity, acceleration, orientation, and speed. External sensors are used to collect information about the external environment surrounding for robot's applications like navigation of robot, object handling, and identification.

7.3.8.3.1 Contact Sensors

Contact sensors have physical contact with the object to be sensed. Commonly used contact sensors are as follows.

Touch Sensor
Touch sensors are used to indicate whether a contact has been established with an object or not. Also, these sensors indicate the presence of the object within the fingers of the end effector. Humans have skin that act as a contact sensor. When one touches a particular object or a material, one understands its physical properties like shape, texture, and temperature. Analogous to the human skin is what is known as tactile sensors in robots. Figure 7.10 shows a typical arrangement of tactile sensors on a gripper and layers of it.

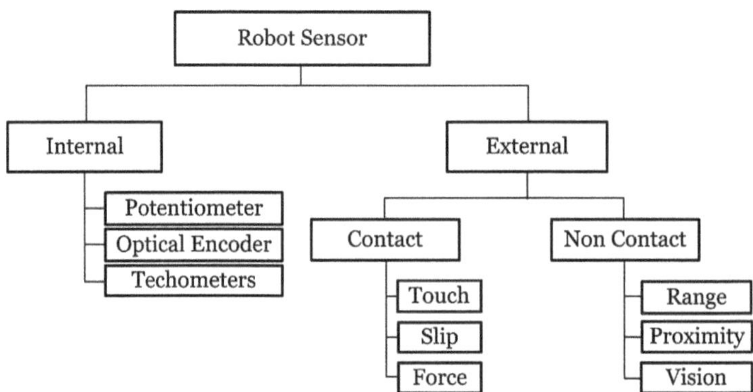

Fig. 7.9 Commonly used sensors in robotics

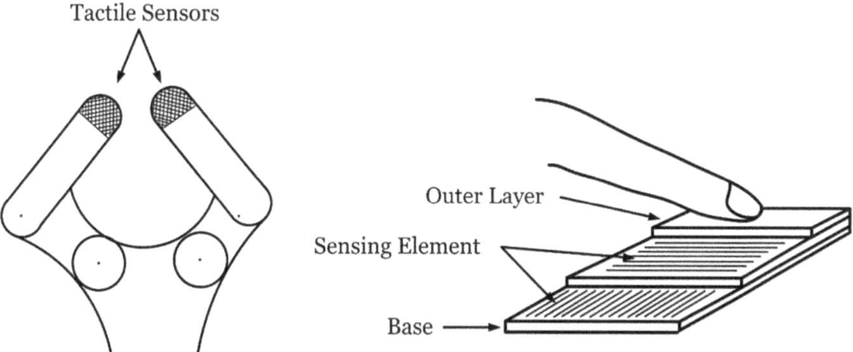

Fig. 7.10 Tactile sensors mounted on a gripper and layers of tactile sensor

Fig. 7.11 Schematic of slip sensor

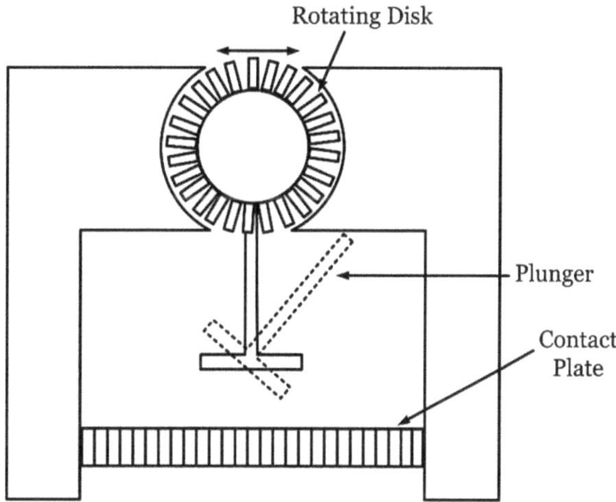

Slip Sensor

Slip sensors are used to specify the contact made by the external objects and determine the slip of the object over the robot body.

Figure 7.11 shows a schematic of a slip sensor. When the object comes in contact tangentially to the rotating disc, the disc rotates. This leads to the change in the contact position of the plunger with the resistive contact plate. This causes changes in resistance across the plunger and contact plate indicating slipping of the object.

Force Sensor

Force sensors are commonly used in a robot's end effector for measuring the reaction forces developed at the interfaces of mechanical assemblies. Most of the force sensors work on the principle of change in resistance or capacitance due to the deflection in the sensing membrane while applying a force. Figure 7.12 shows a robot gripper with force sensor.

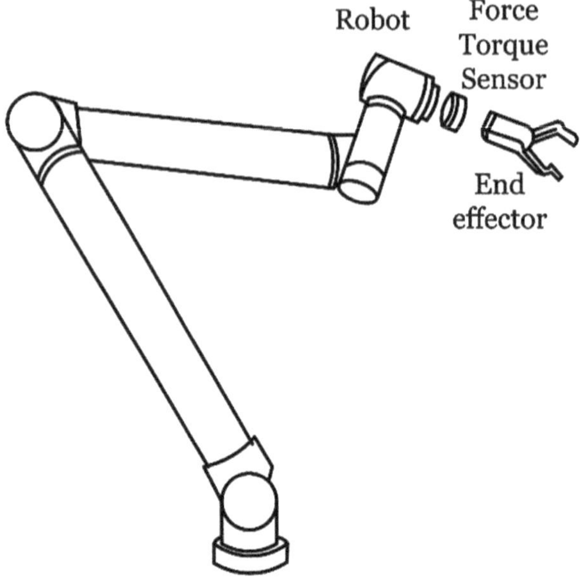

Fig. 7.12 Force sensor mounted on the gripper of a robot

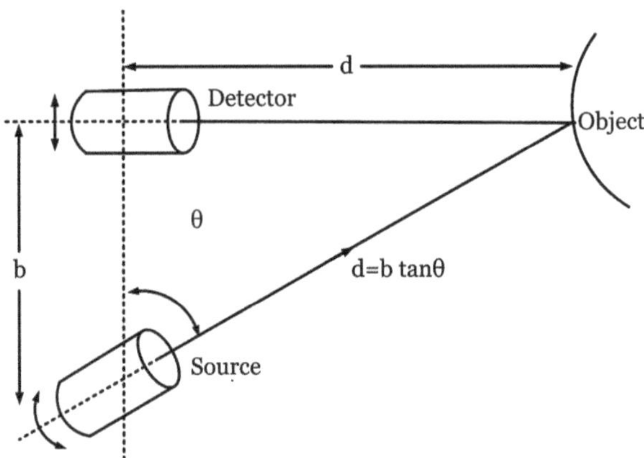

Fig. 7.13 Schematic of working of range sensor

7.3.8.3.2   Non-contact Sensors

Non-contact sensors have no physical contact with the object or surroundings. Commonly used non-contact sensors are as follows.

Range Sensor

Range sensor is mainly used to measure the distance between a reference point on the robot and the objects present in its workspace. These sensors are primarily based on the working principle of optics. In Fig. 7.13, d is the distance between a reference point on the robot and the object, and is given by

$$d = b\tan\theta$$

where b is the distance between the detector and the source of the sensor, and θ is the angle between the source and the plane perpendicular to detector. The parameter b is known for specific sensors and θ is estimated using the principles of optics. Using b and θ, the distance of the object from the robot is calculated.

Proximity Sensor

Proximity sensors are used to indicate the presence of an object within a specified distance/interval without having a physical contact with the object. When an object comes near the circuit, due to the presence of high-frequency magnetic field, deflection occurs in the internal circuit which sends the signal to the controller, and thus it opens or closes or actuates the actuator. One can set a particular target to the proximity sensor, as shown in Fig. 7.14. Generally, this type of sensors has a proximity switch and an active face. As soon as the target comes to the particular range, the active face sends signal to the proximity switch which sends the signal to the controller. This is generally a photoelectric sensor which can be installed in the robot to detect the presence of an object in its work volume.

Vision Sensor

Vision sensors are used to recognize three-dimensional objects in the form of 3D/ 2D images or in two dimensions. These sensors are also called robot vision, machine vision, or artificial vision sensors (Fig. 7.15).

This can be used to detect the good or bad products in an inspection department of a manufacturing company. A vVision sensor first inspects a part and compares it with an image that was already fed to the computer. The outcome of the process informs whether the object has passed or failed the quality test.

7.3.9   *Intelligent Control Systems in Robotics*

Robots are employed for both industrial and personal purposes. Most robots in industries follow planned procedures to carry out their operations and are not accustomed to deal with unexpected or unplanned situations. Traditional techniques of controlling a robot such as PID controllers have been used over the years, but they can perform efficiently in limited environmental conditions only. With an increase in operational environment complexity, robots are expected to perform self-diagnosis, self-reorganization, and detect fault tolerance on their own. These features need decision-making capability by the robot leading to the requirement of intelligent control systems. An intelligent control system enables a robot to learn how to respond to unknown dynamic environment or situations. Intelligent control systems are generally achieved using advanced software computations integrated with sensors. A schematic representation of intelligent techniques from the domain of artificial intelligence for robotics control is presented in Fig. 7.16.

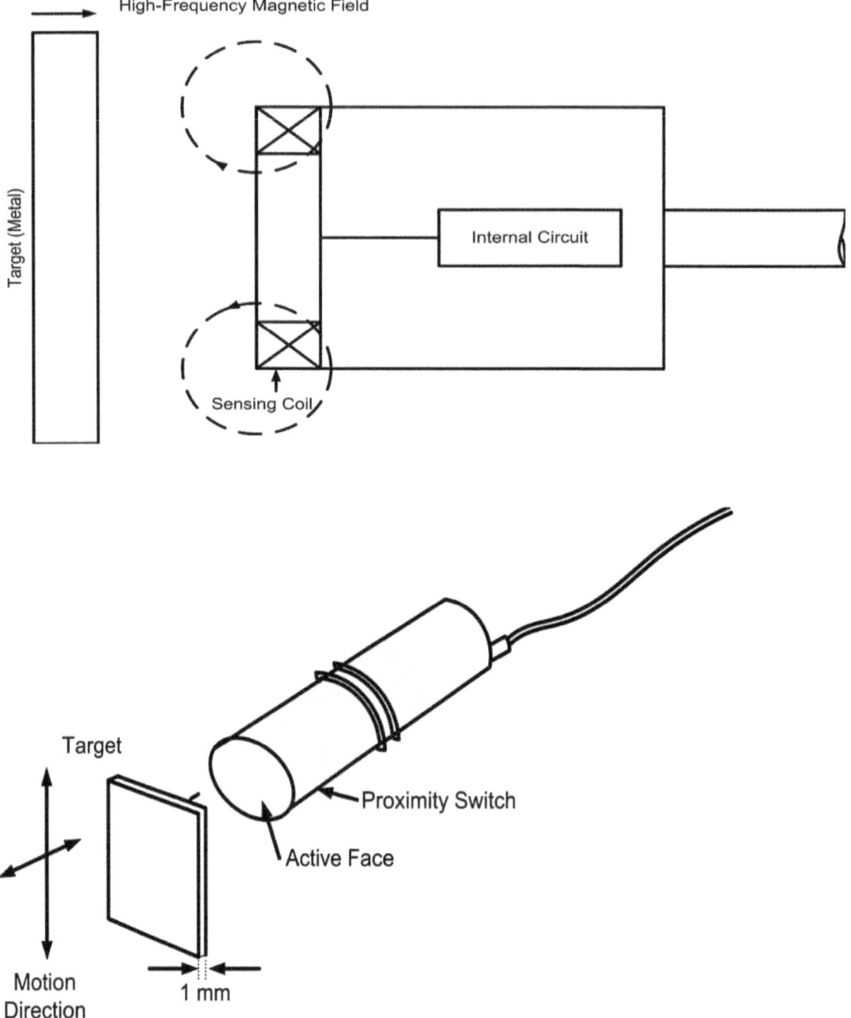

Fig. 7.14 Proximity sensor

Methods for Providing Intelligence to Control Systems
Fuzzy logic and artificial neural networks are among the important methods that can provide intelligence to a robotic control system.

7.3.9.1   Fuzzy Logic

Fuzzy logic is a method of soft computing based on degrees of truth as opposed to only a true or a false (1 or 0) decision. A fuzzy control system for behavior-based

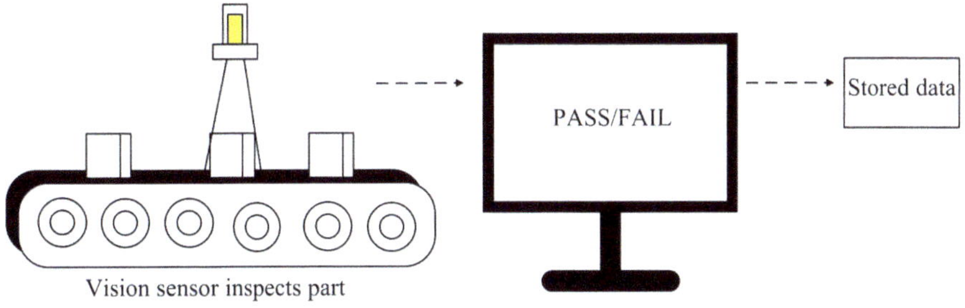

Fig. 7.15 Vision sensor

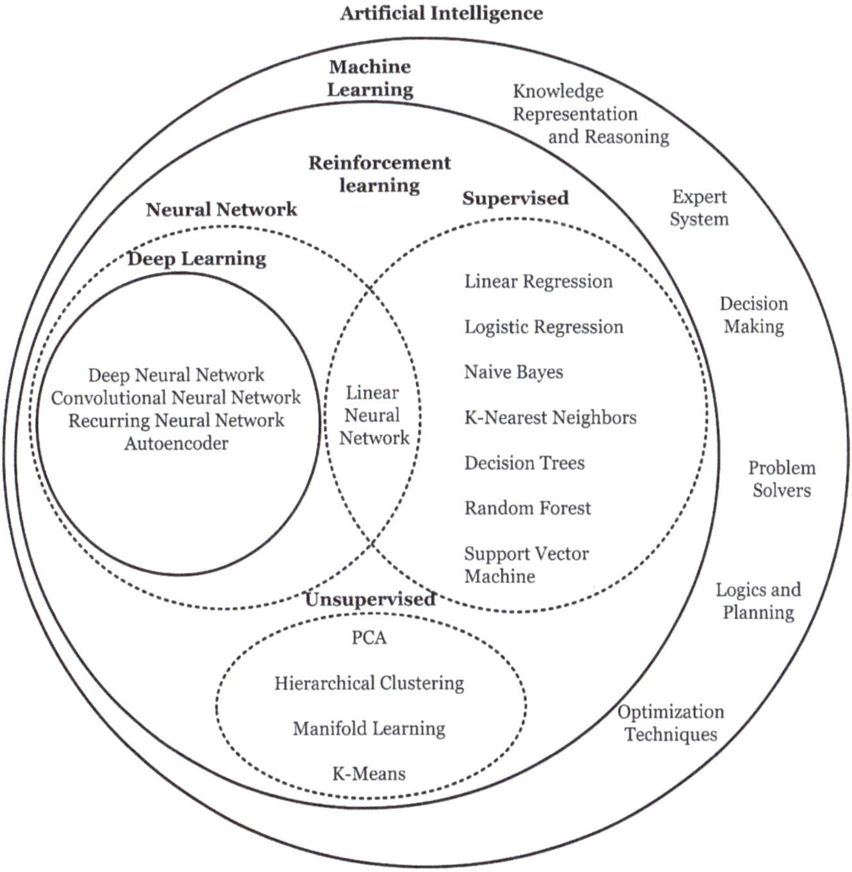

Fig. 7.16 Representation of techniques used for intelligent decision-making in robots

robotic system would start with crisp sensor readings (e.g., numeric values from proximity sensors), translating them into linguistic classes in the fuzzifier, firing appropriate rules in the fuzzy inference engine, and generating a fuzzy output value. The fuzzy output value is further translated into a crisp value that represents actuator control signals. Fuzzy logic allows a certain type of discrete encoding of the fuzzy input by using rule-based systems.

Base rules are represented as a collection of if–then rules which take the general form:

$$if\ antecedent\ then\ consequent$$

where the antecedent consists of a list of preconditions which must be satisfied in order for the rule to be applicable and the consequent contains the output response. The fuzzy logic controller (FLC) that depends on fuzzy logic gives a method for adjusting a linguistic control methodology in view of expert information into the programmed control strategy.

7.3.9.2 Artificial Neural Networks (ANNs)

Neural networks are computing systems that mimic the functioning of the human brain. It is based on collection of connected nodes called artificial neurons, like neurons present in the human brain. Like biological neurons that communicate with one another through synapses, artificial neurons also receive, process, and pass on the signals to one another. Artificial Neural Networks (ANNs) attempt to emulate the biological neural network. An artificial neuron which is the basic element of an NN contains three main components: weights, threshold, and activation function.

Chapter 8
Academic Projects and Tools in Robotics

This chapter presents projects on robotics and embedded systems demonstrating the implementation of the concepts discussed in the previous chapters. It is envisaged that these projects will encourage readers for undertaking innovative projects following a systematic approach for realizing textbook concepts through real-world visualization and thereby attempting to solve societal issues. The tools and equipment usually used while doing robotics are briefed in this chapter.

8.1 Developement of a Cost-Effective EMG-Controlled Three-Fingered Robotic Hand

The objective of this project is to enable students to acquire skills in terms of creative thinking, time management, and self-learning while earning academic credentials. This project was awarded the first position in IEEE Best Students' Project Award during March 1–2, 2012, held in Maulana Azad National Institute of Technology, Bhopal.

8.1.1 Project Planning

The implementation of the project requires a systemic plan comprising of the following milestones:

1. Study on the requirement of the project and its importance
2. Review of the available literature in terms of research and commercial prototypes
3. Timeline for target activities vis-à-vis tasks to be completed for the project

N. M. Kakoty et al., *Introduction to Embedded Systems and Robotics*,
https://doi.org/10.1007/978-3-031-73098-6_8

4. Contingency plan to deal with situations that could not be seen while planning for the project
5. Documentation and presentation of the project

8.1.2 Overview of Technical Details

According to a 2011 report by Mediescapes India, it was estimated that 2,20,000 of people in India are upper limb amputees. Rehabilitation of these people using prosthetic limbs would provide them assistance in their day-to-day lives. The design and development of prosthetic hands pose many challenges with respect to flexibility of the robotic hand, finger movement, and grasping adaptability. Although many research works were reported until 2011 in the field of robotic hand to eliminate these problems, yet many more challenges were still needed to be addressed. This academic project implemented during 2011–2012 addressed one such major problem encountered in the development of robotic hand prostheses, that is, the achievement of electromyogram (EMG)-based control with sensory feedback.

EMG is a diagnostic recording at the surface of the skin for analyzing the electrical activity produced by the skeletal muscles. The research took its lead by the fact that despite the available commercial hand prosthetics with advanced technology, they were far from the original level of grasp self-adaptibility with regard to the object to be grasped.

The project implemented localized learning as the learning model to develop a three-fingered hand with self-adaptibility according to the shape of the object to be grasped. The three-fingered robotic hand possessed six degrees of freedom using three actuators, where each actuator was responsible for providing one active and one passive degree of freedom. The hand could grasp oval, cuboid, circular, and cylindrical objects. A grasp planing controller based on EMG signals was made to command for actuating the hand opening and closing. A fuzzy classifier in the grasp planner was used for recognizing the shoulder abduction and adduction movement based on root mean value of EMG. The control was through a proportional controller customized with position and touch sensors. The sensory feedback was in order to adapt the hand to the shape of the object to be grasped. Figure 8.1 shows the developed EMG-controlled three-fingered hand.

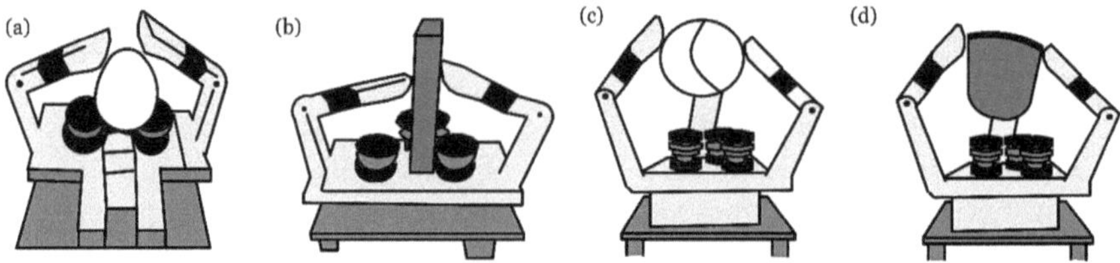

Fig. 8.1 Developed hand grasping four types of objects (**a**) oval (egg), (**b**) cuboid (mobile phone), (**c**) circular (cricket ball), (**d**) cylindrical (cup)

A research paper entitled "Development of cost-effective EMG-controlled three-fingered robotic hand" based on control system involved in this work was presented in the 3rd IEEE International Conference on Computer and Communication Technology 2012 held in Motilal Nehru National Institute of Technology, Allahabad. Working video of this project is available at "https://www.youtube.com/watch?v=3 dqSRxO5E2M" titled *A Three Fingered Hand Customized with Position and Touch Sensors.*

8.2   Development of a Biomimetic Prosthetic Finger

This is a graduate-level student project with an objective to design and develop a biomimetic prosthetic finger. Biomimesis is the study of biological systems, its models, and processes and their emulation in artificial systems. The planning for the project is in line with the project in Sect. 8.1. Well-defined planning and the efforts of the student team along with Professor S. K. Mukharjee are appreciated, who:

1. Won GOLD Medal in Anveshan 2013, a student research convention organized by the Association of Indian Universities, March 20–22, 2013, held in Tata Institute of Social Science, Mumbai
2. Won first position in IEEE Best Student Project Award 2013, March 2–3, 2013, held in Maulana Azad National Institute of Technology, Bhopal.

The project was also presented as a research paper entitled *"Development of a biomimetic prosthetic finger"* in the International Conference on Advances in Robotics 2013 held in Defence Research Development Establishment, Pune.

8.2.1   Overview of Technical Details

A human hand helps interact with the environment in our day-to-day activities. Finger amputation is one of the most frequently confronted forms of partial hand losses. Despite serious research works, little progress has been made toward a functional prosthetic finger. Low functionality with respect to degrees of freedom (DoF), joint range of motion (RoM), and unrelated actuation mechanism are the main reasons for non-acceptance of finger prosthesis by amputees. Furthermore, available market variants were far from the human finger in terms of deploying abduction-adduction movement.

This project presented the development of a prosthetic finger prototype following a biomimetic approach inspired by human finger anatomy and physiology. The developed protoype mimicked the human finger in dimensions, joint RoM, DoF, and electromyogram (EMG)-based control mechanism. It could perform both flexion-extension and abduction-adduction movements following the EMG recognition

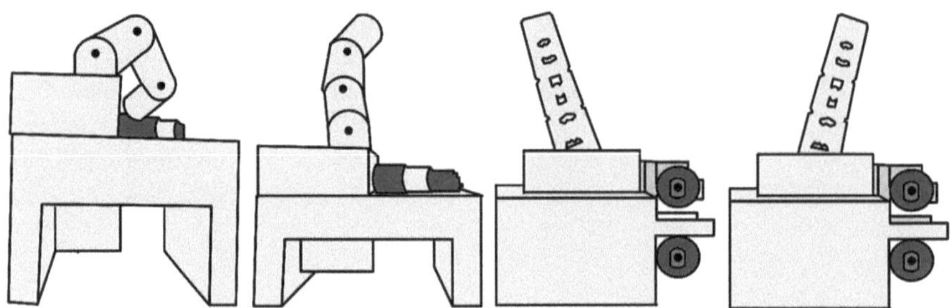

Fig. 8.2 Prosthetic finger prototype performing flexion-extension and abduction-adduction movements

for the intended type of movements. The torque from the actuation unit was transmitted to the finger joints through an antagonistic tendon mechanism. The biomemetic approach followed for the design and development was categorized into three main stages: (i) detailed study of the human finger anatomy and physiology, (ii) derivation of useful functions and processes like development of computer aided design (CAD) finger prototype, selection of material to be used as skeletal, choice of actuation and tendon mechanism as well as static and dynamic constraints, and (iii) imitation of the derived functions and processes in artificial systems, i.e., into the prototype shown in Fig. 8.2. The working video of this project is available at "https://www.youtube.com/watch?v=P9Hoo2VvEG8".

8.3   A Mobile Robot for Hazardous Gas Sensing

Realizing the challenges in military war field, an ambitious project on a mobile robot equipped with hazardous gas sensing capability was carried out during 2018–2020 at a graduate level. The project involved development of a mobile robot prototype customized with a module of gas sensors, human detection sensor, GPS, and obstacle detection sensor with wireless monitoring system. This project was nominated as one of the 21 innovative projects in the e-Yantra Ideas Competition 2019 organized by the Indian Institute of Technology Bombay. This project was then published as a research paper entitled *"A Mobile Robot for Hazardous Gas Sensing"* in IEEE International Conference on Computational Performance Evaluation, 2020.

8.3.1   Project Planning

First, the reliable components to be used and the tasks to be followed are identified:

Task 1: Designing and development of the mobile robot

> Sub-task 1: CAD model of the robot
> Sub-task 2: Physical model of the robot
> Sub-task 3: Mounting of the actuators (DC and servo motors) and sensors (ultrasonic, PIR, gas sensor)
> Sub-task 4: Mounting of the GPS module, ZigBee module

Task 2: Testing of the actuators and sensors mounted in Task 1

> Sub-task 1: Testing of DC and servo motors
> Sub-task 2: Testing of the ultrasonic, PIR, gas sensors
> Sub-task 3: Testing of the GPS and ZigBee module

Task 3: Development of a navigation algorithm for unknown dynamic environment

> Sub-task 1: Development of the flowchart for the navigation algorithm
> Sub-task 2: Development of the code for the navigation algorithm
> Sub-task 3: Testing of the algorithm in unknown dynamic environment

Task 4: Transmission of information to the soldier's site from the robot

> Sub-task 1: Interfacing of the gas sensor module to the ZigBee module for transmission of gas information
> Sub-task 2: Interfacing of the GPS module to the ZigBee module for transmission of robot's location

Table 8.1 shows the bar/gantt chart or time frame of the project.

8.3.2   Overview of Technical Details

The developed mobile robot shown in Fig. 8.3 is based on a six-wheeled rocker-bogie mechanism. This mechanism with wheel formula (total number of wheels (6) × number of actuated wheels (6) × number of steerable actuated wheels (4) = 144) allows the robot to navigate in uneven terrain and rotate 360 degree at zero radius. The robot can sense the presence of hazardous gases in the environment and map the locations of detected gases in real-time using GPS.

The prototype robot was tested for recognizing hazardous gases like carbon dioxide, liquefied petroleum gas, and vaporized alcohol gas. A handheld control unit was developed to have manual control of robot navigation and, receive information about the type and concentration of the gases under study, presence of human beings in proximity to the robot and the GPS location of the robot. The designed robot could avoid collision with obstacles while navigating using ultrasonic sensors. A neural network-based classifier was implemented to recognize the gases with an average accuracy of 98%. All the communication between the robot and the hand-held control device was accomplished using Zigbee modules. Working video of this project is available at "https://www.youtube.com/watch?v=jDRiWCqq5Yo".

Table 8.1 Gantt chart

	Time frame					
	0–1 Month	1–2 Month	2–3 Month	3–4 Month	4–5 Month	5–6 Month
Task 1: Sub-task 1	■					
Task 1: Sub- task 2	■					
Task 1: Sub-task 3	■					
Task 1: Sub-task 4		■				
Task 2: Sub-task 1		■				
Task 2: Sub-task 2			■			
Task 2: Sub-task 3				■		
Task 3: Sub-task 1					■	
Task 3: Sub-task 2					■	
Task 3: Sub-task 3					■	
Task 4: Sub-task 1						■
Task 4: Sub-task 2						■

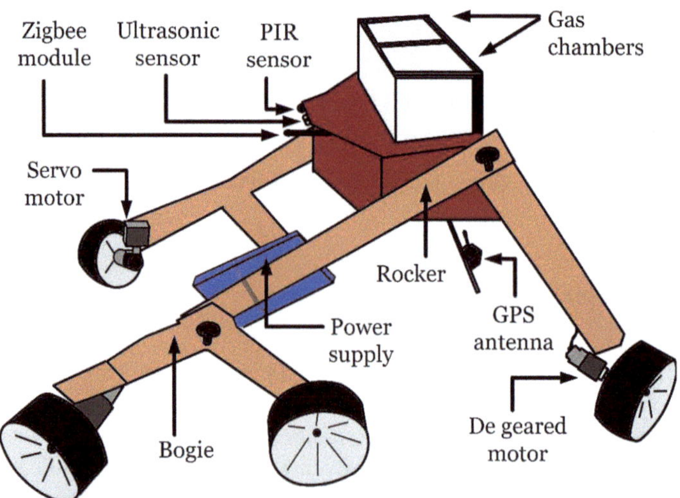

Fig. 8.3 Mobile robot with its functional units

8.4 A Real-Time EMG-Based Prosthetic Hand with Human-Like Grasping

Development of prosthetic hands with human-like functionality and controllability is one of the major goals in the area of rehabilitation robotics. Current developments on prosthetic hands have earned higher functionality with multiple fingers and degrees of freedom. However, the time required to perform a grasp-type operation opens avenues for improvement in its controllability. In order to address this issue, this project executed by one of the postgraduate students focused on the development of a real-time EMG-based prosthetic hand with human-like grasping time. The results of this project have been published with a patent application number 201931049269 dated June 4, 2021. The output of this project was published in the journals *NeuroComputing* and *Intelligent Robotics and Applications*.

8.4.1   Overview of Technical Details

Figure 8.4 shows the experimental setup of EMG-based prosthetic hand control. EMG-based prosthetic hand control architecture was implemented in three phases. In the first phase, the prosthetic hand initially remained in open state. Raw EMG data were collected from the biceps branchii muscles of the subjects using an EMG unit through Ag/AgCl electrodes. The instrumentation amplifier amplifies EMG with a gain of 100 and common mode rejection ratio (CMRR) of 110 dB. The acquired EMG was passed through a filter with pass band of 10–500 Hz. In the second phase, the pre-processed EMG was digitized with a 10-bit analog to digital converter (ADC) at 8.2 KHz. The digitized EMG was processed in the controller wherein windowing, feature extraction, and recognition of user's intention for grasping were accomplished. EMG was segmented with a 50 milli-second window. Root mean square (RMS), a time-domain feature, has been used to quantify the EMG. RMS was chosen as feature because it reflects the physiological activity in the motor unit during contraction. A finite state algorithm (FSA) was implemented to understand the user's intention for grasping operation. The choice of FSA was based on the fact that it involved lesser computational complexity as compared to the machine

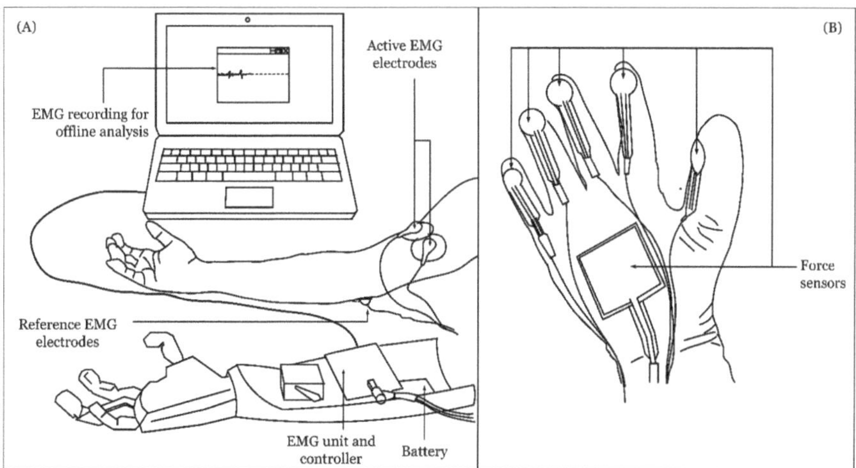

Fig. 8.4   (**a**) Experimental setup of EMG-based prosthetic hand control. (**b**) Data glove equipped with force sensors

learning algorithms and thereby was suited for real-time embedded applications. In the third phase, on recognizing user's intention for grasping, the controller commands the actuators for grasping by the prosthetic hand. Experiments have been accomplished in four sessions, each with 20 trials, by five subjects in both sitting and standing positions. It has been found that the prosthetic hand can perform grasping with an average accuracy of $96.2 \pm 2.6\%$. The controller satisfied the neuromuscular constraint enabling the prosthetic hand to perform grasping operation in 250.80 ± 1.1 milliseconds, which is comparable to the time required by human hands, i.e., 300 milliseconds.

8.4.2   Experimental Results

Figure 8.5 shows grasping of objects in real time by the prosthetic hand. Real-time experiments with the embedded EMG controller were performed by the subjects in four sessions for grasping the four objects under study. During this, subjects were either in sitting or standing positions randomly. The grasping performances have been evaluated in terms of grasping accuracy, success to EMG intensity ratio, and grasping time.

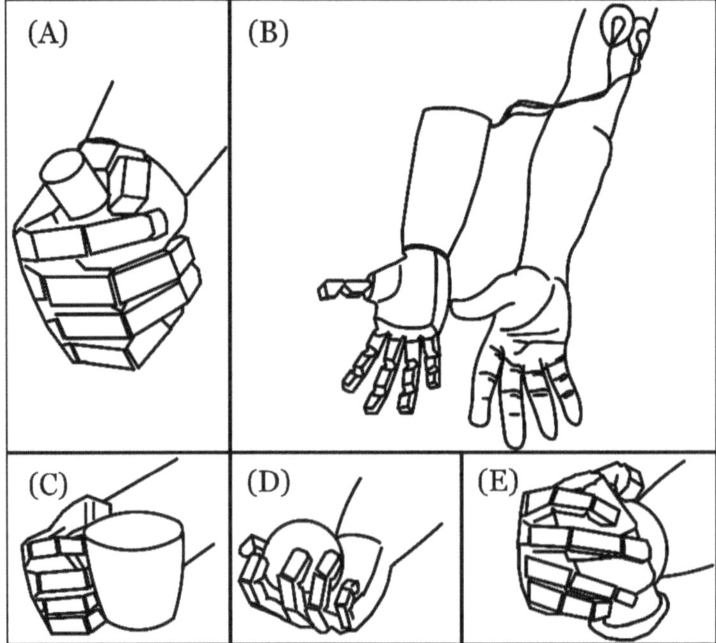

Fig. 8.5 Prosthetic hand grasping the objects in real time (**a**) prosthetic hand grasping a plastic container, (**b**) subject wearing the prosthetic hand, (**c**) prosthetic hand grasping a coffee mug, (**d**) prosthetic hand grasping a cricket ball, (**e**) prosthetic hand grasping a screw-driver box

8.4.2.1   Grasping Accuracy

It is qualitatively defined as the ability to perform grasping task by a prosthetic hand without dropping the objects. It indicates the grasping capability of the prosthetic hand and possible acceptability by the user. In this study, the user actuates the prosthetic hand by EMG generated through maximum voluntary contraction (MVC) of biceps muscles for grasping objects. The number of times that the prosthetic hand can grasp without dropping the grasped objects during the experiment was recorded. Therefore, it is expressed as the grasping accuracy (A) and defined as in Eq. 8.1.

$$A = \frac{\text{Correctly Grasping Instances}}{\text{Total Number of Grasping Instances}} \times 100\% \tag{8.1}$$

Figure 8.6 shows the grasping accuracy average across the subjects while trying to grasp the four objects. It has been found that the subjects could use the prosthetic hand with an average grasping accuracy of 96.8 $\pm$ 3% in sitting position and 95.6 $\pm$ 2.2% in standing position.

8.4.2.2   Success to EMG Intensity Ratio

Success to EMG intensity ratio is the percentage of successful grasping to average EMG intensity generated by a user across four sessions. It is defined as P in Eq. 8.2.

$$P = \frac{\sum_i^N A(i)\%}{I_{\text{EMG}}} \tag{8.2}$$

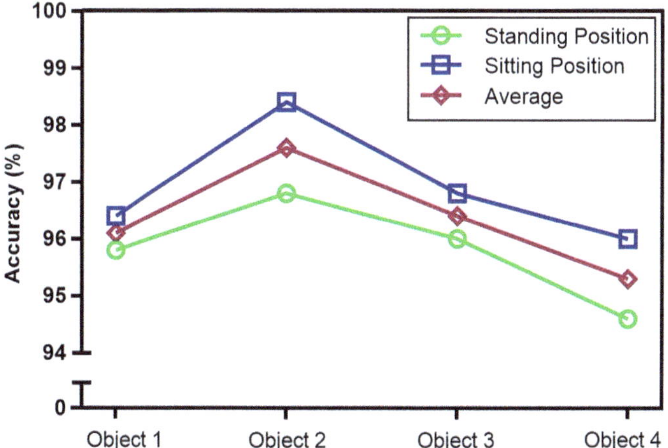

Fig. 8.6 Grasping accuracy for the four objects averaged across all subjects

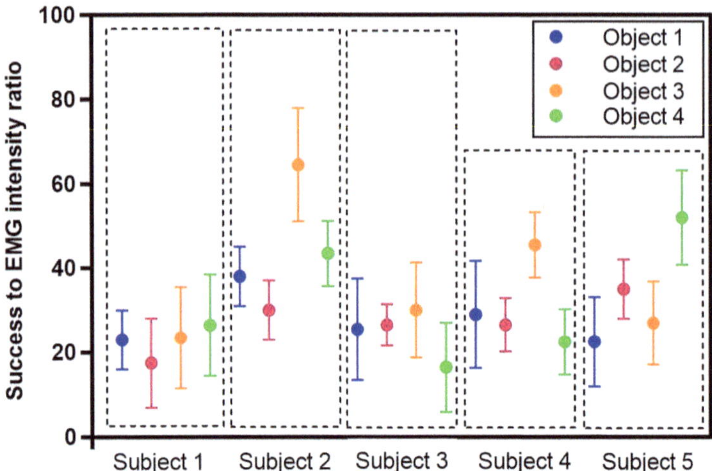

Fig. 8.7 Success to EMG intensity ratio of the five subjects

where N represents the total number of sessions. The average EMG intensity represented as IEMG is defined as in Eq. 8.3.

$$I_{\text{EMG}} = \frac{1}{N} \sum_{i=1}^{T} \frac{e(i) - e_{\min}}{e_{\max} - e_{\min}} \tag{8.3}$$

where $e(i)$ is the ith sample of EMG, T represents the total number of trials during the experiment, emin and emax are minimum and maximum value of EMG, respectively. Figure 8.7 shows the success to EMG intensity ratio for five subjects across four sessions of 20 trials each while grasping object by the prosthetic hand. Success refers to the grasping accuracy following Eq. 8.1. EMG intensity depicts the subjects' effort to prevent dropping of object during grasping supported with visual feedback. Observing the variation of the success to EMG intensity ratio compared to the average accuracy shown in Fig. 8.6, it could be understood that EMG intensity or users' effort was made to prevent dropping of different objects.

Two-way analysis of variance (ANOVA) was performed for perusal of this variation. Table 8.2 shows the results of ANOVA of grasping accuracy vis-à-vis success to EMG intensity ratio. The significant p-value indicates a statistically significant variation between grasping accuracy and success to EMG intensity for each subject. This signifies that the users have put different efforts in terms of EMG intensity for maintaining a stable grasp without dropping the objects and thereby ensuring a good grasping accuracy through visual bio-feedback.

Table 8.2 ANOVA of grasping accuracy vis-à-vis success to EMG intensity

Subjects	Sum of squares	df	F-statistic	p-value
1	10911.40	1	994.13	0.0001
2	5559.94	1	49.45	0.0059
3	10746.10	1	605.80	0.0001
4	8711.84	1	171.25	0.0010
5	8115.18	1	80.56	0.0029

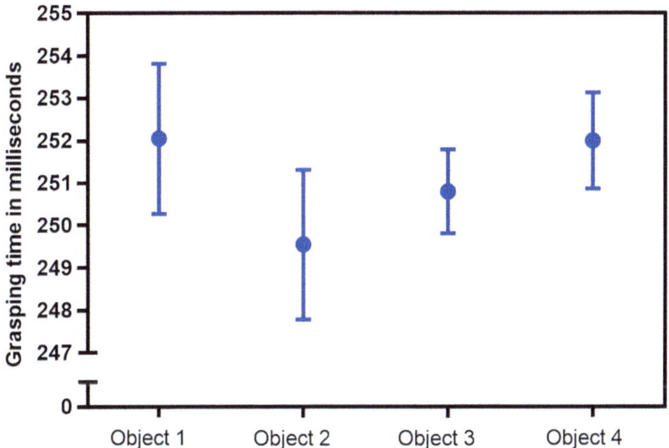

Fig. 8.8 Grasping time from initiation of EMG to grasping objects

8.4.2.3   Grasping Time Estimation

The grasping time required by the prosthetic hand was estimated using the data glove shown in Fig. 8.4 while grasping the four objects. Six force sensors were equipped in the data glove each for measuring the contact forces of the five fingers, i.e., thumb, index, middle, ring, and little along with the palm. Another controller was used externally to read the electrical signals from the force sensors for indicating a stable grasping of an object. A minimum force of 3N in any two or more sensors signifies a successful grasp close action by the prosthetic hand. The time elapsed method was utilized to record the time instants of EMG received by the EMG unit for a grasping action and the stable grasping of the object by the prosthetic hand in real time. The time difference hence calculated serves as the grasping time by the prosthetic hand controller. Figure 8.8 represents the estimated grasping time by the prosthetic hand calculated using the data glove for the four objects. The grasping times were estimated to be 251.9 ± 1.1 milliseconds, 248.8 ± 1.4 milliseconds, 250.9 ± 0.9 milliseconds, and 251.7 ± 0.8 milliseconds corresponding to the four objects. The typical grasping time by the prosthetic hand controller for any object was calculated as 250.8 ± 1.1 milliseconds. The working video of this project is available at "https://www.youtube.com/watch?v=22JZ4rmPoCc".

8.5 A Real-Time EMG-Based Prosthetic Hand with Multiple Grasping

The human hand performs multiple grasp types during daily living activities. Adaptation of grasping force to avoid object slippage as employed by human brain has been postulated as an intelligent approach. Recently, research for prosthetic hands with human-like capabilities has been followed by many researchers, but with limited success. Advanced prosthetic hands that can perform different grasp types use multiple EMG channels. This causes the user to wear a greater number of electrodes leading to inconveniences and with inadequate grasping accuracy by the prosthetic hands.

This project reports a prosthetic hand performing 16 grasp types in real time using a single-channel EMG customized with an Android application Graspy. An embedded EMG-based grasping controller with a network of force-sensing resistors and kinematic sensors prevent slipping and breaking of grasping objects. Experiments were conducted with four able-bodied subjects for performing 16 grasp types and individual finger movements. A proportional-integral-derivative (PID) algorithm was implemented to regulate kinematic constraints of the prosthetic hand fingers in linear relation to the force sensing resistors. The control algorithm can prevent slipping and breaking of grasping objects with a finger joint angle reproduction precision of $0.16°$. The hand could perform grasping of objects like tennis ball, cookie, knife, screw-driver, water bottle, egg, pen, and plastic container, while emulating the 16-grasp types with an accuracy of 100%. The experimental setup is shown in Fig. 8.9. The hand grasping four objects is shown in Fig. 8.10.

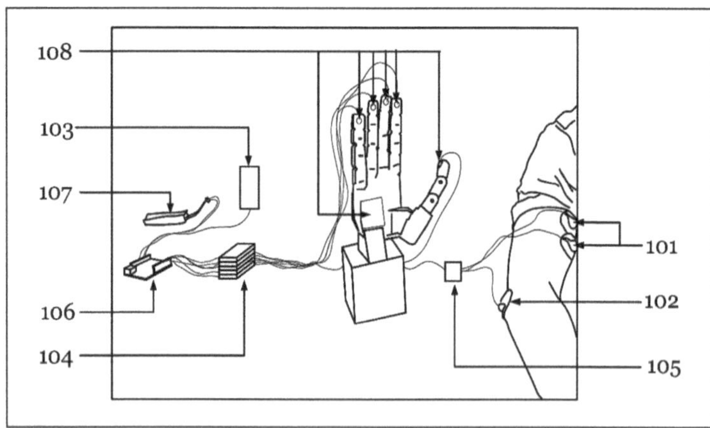

Fig. 8.9 Experimental setup: 101: active EMG electrodes; 102: reference EMG electrode; 103: wireless receiver; 104: PID controllers; 105: instrumentation amplifier and band pass filter; 106: microcontroller; 107: electrical power source (3300 mAh); and 108: force sensors

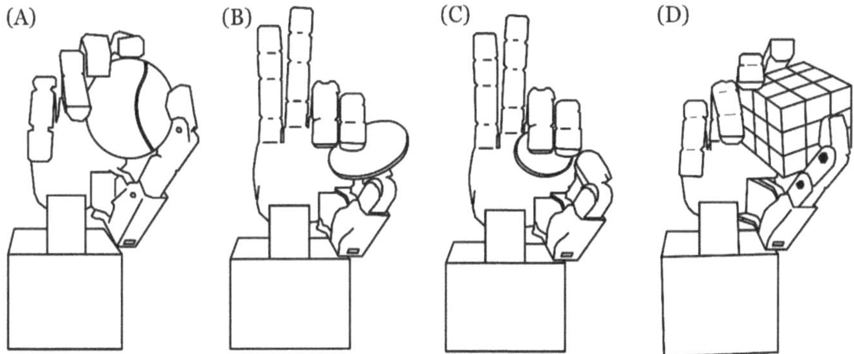

Fig. 8.10. Prosthetic hand grasping objects like tennis ball (**a**), cookie (**b**), egg (**c**), and Rubik's cube (**d**).

Experiments were conducted on the developed prosthetic hand with four subjects for testing multiple grasp performance accuracy and adaptive capability of the prosthetic hand.

Figure 8.11 shows the time-line parameters associated with prosthetic hand control architecture. Figure 8.11a shows RMS values of real-time pre-processed EMG in millivolts containing two events of MVC by the user. Simultaneously, Fig. 8.11b, c reflects individual fingertip forces on the grasped object and MCP joint angles of the individual fingers. It can be seen from this figure that while performing a grasp type, finger joint angles were adjusted depending upon the force on their respective fingertip. Figure 8.11d, e displays instances of grasp close and grasp open by the prosthetic hand.

8.6 Tools and Equipment

This section states different types of mechanical, electrical, electronics components, software tools, and equipment used in the development of robotic platforms.

8.6.1 Mechanical Tools

Some of the important mechanical tools required while working on a project in robotics are:

1. *Measuring tape*: A measuring tape is a flexible ruler available in both inches and metric marking systems. It is a common measuring tool in robotics used to measure distances and dimensions of objects.

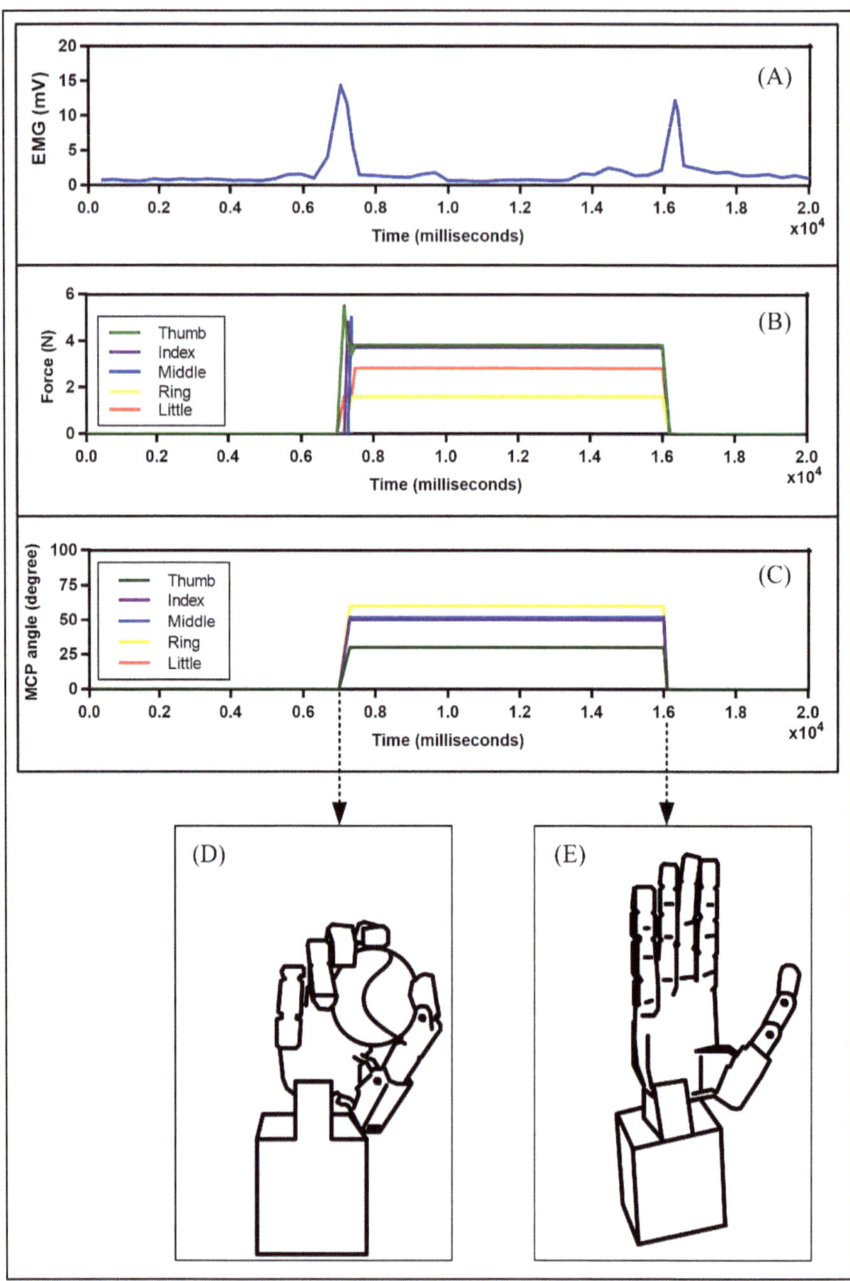

Fig. 8.11 (**a**) EMG from bicep brachii muscle actuating the grasping presented in (**d**), (**b**) force generated at fingertips during the grasping presented in (**d**), (**c**) MCP joint angles of the five fingers during grasping presented in (**d**), (**d**) grasping by the prosthetic hand (**e**), grasp open by the prosthetic hand actuated by the EMG

2. *Table vice*: A table vice is a tool used for holding objects to facilitates cutting and drilling operations on it. Objects of different sizes can be held firmly by adjusting the vice jaw through a screw mechanism.
3. *Screwdriver*: A screwdriver is used to carry out screwing of miniaturized, small, and medium fastenings or loosening of screws of different sizes. It can be either electrically powered or manually operated.
4. *Wrenches*: Wrenches are used to tighten or fasten bolts and nuts of various sizes.
5. *Saw*: It is used for cutting and is a very important tool for building a robot. The common blade size ranges from 10 to 12 inches in length.
6. *Vernier caliper*: A Vernier caliper allows marking out and measurement up to a precision of 0.01 mm.
7. *File*: A file is used for smoothening the rough edges of a work-piece. Different types of files like round, half-round, and flat are available for specific applications.
8. *Centre punch*: A center punch is used for accurate marking of holes to be drilled.
9. *Drill press*: A drill press is a tool that is used for accurate drilling of the marked holes. It can either be electrically powered or manually operated.
10. *Utility knives*: Utility knives are also called carpet knives used to cut plastic, rubber, paper, and other soft materials.
11. *Hot glue guns*: Hot glue guns are used for gluing parts quickly. They are used in a number of applications like attaching two work-pieces for properly routing the electrical connectors and protect the circuits from damage by exposure to moisture and water.
12. *Safety goggles*: Safety goggles are an important robotic tool used to protect the eyes as the fine particles that come out abruptly while working on a machine tool are dangerous to the eyes.

8.6.2 *Electrical and Electronic Tools*

Electrical and electronic tools are used for developing the power supply and control units of robotic systems such as driver, controller, and buffer circuits. Some of the commonly used electrical and electronic tools are:

1. *Soldering iron*: A soldering iron is used for all routine soldering of electronic and electrical parts. Anti-static wrist strap is a commonly used tool while using soldering iron to prevent the buildup of static electricity near sensitive electronics components.
2. *Soldering gun*: A soldering gun is a pistol-shaped, electrically powered tool used for soldering metal work pieces. It uses a tin-based solder to build a strong mechanical bond.
3. *Bench power supply*: A bench power supply is used to supply consistent voltage to provide biasing voltage to a circuit during development and testing of robots.
4. *Sensors*: Sensors are used for converting a physical phenomenon to equivalent electrical signal. Different types of sensors used in robotics are light sensors,

sound sensors, temperature sensors, contact/touch sensors, proximity sensors, pressure sensors, and tilt/acceleration sensors.

5. *Integrated circuit*: An integrated circuit is an assembly of discrete components like diodes, transistors, resistors, capacitors fabricated onto a wafer in a single package. The commonly used integrated circuits in robotics are voltage regulators, motor driver circuits, and microcontroller circuits.

6. *Actuator*: An actuator is an electromechanical device that converts electrical energy into mechanical energy. It is used to bring the changes in the physical environment according to the signal from the controller with inputs from sensors. Different types of commonly used actuators are electrical, pneumatic, hydraulic, piezoelectric, electromechanical, and shape memory alloy.

8.6.3   Software Tools

Software tools are used for design and analysis of robotic systems. It is very useful as the behavior and functioning of a robotic system can be evaluated before testing with a physical system. Further, these tools are used in teaching insights of the functioning of robotic principles. Some of the commonly used software tools are:

1. *RoboAnalyzer*: It is a licensed software used for teaching and learning robotic concepts. It is a 3D model-based software that uses a virtual platform for modeling the robots. In this tool, uses of translational and rotational matrices, Denavit-Hertenberg parameters, kinematic and dynamic analysis can be realized and visualized in practice. It has the feature of integrating the modeled robots to various other software tools such as MATLAB, MS-Excel, and other applications via a COM interface. This toolbox along with tutorials is available at www.roboanalyzer.com.

2. *Peter Corke's robotics toolbox*: It is a MATLAB-based robotics toolbox. The toolbox contains functions and classes to represent orientation and pose in as matrices. It uses MATLAB functions for representing kinematics and dynamics of serial-link manipulators. The tool box is available at https://petercorke.com/toolboxes/robotics-toolbox/.

3. *jmeSim*: It is an open-source, multi-platform robotic simulation tool. This robotic software tool is a JAVA-based robot simulator that provides excellent graphical and physical fidelity. It has integration with Robot Operating System (ROS). It comprises a bundle of sensors such as thermal camera, depth camera, sonar sensor, and laser range finder along with many actuators for simulation of robotic systems. A desired simulation environment can also be created using this tool.

8.6.4 Equipment

With the advances in automation technology, many tiresome tasks involved in the development of robotic systems have become much easier and simpler using equipment. Three such equipment are:

1. *3D printer*: A 3D printer can construct computer-aided 3D model robotic systems or their parts using fused deposition technique. It uses slicing software to convert the modeled 3D structures into geometric codes. These codes on being fed to the 3D printer are interpreted by its controller. The controller commands the actuators and the nozzle to perform the deposition of fused filament layer by layer developing the physical model.
2. *PCB printer*: A PCB printer allows the user to print a circuit diagram developed in a PCB-design software, which, in turn, allows the user to assemble components on the printed circuit board. Development of a printed circuit board using PCB printer is much faster and easier compared to the traditional method. Each PCB printer includes an inkjet print head that can dispense fine droplets of conductive ink and insulating ink, allowing to create multi-layered rigid and flexible circuits on FR-4, Kapton, or any substrate of the developer's choice. In addition, each printer has two heads, one for dispensing glue and solder paste and another to pick-and-place components to be used in the circuit.
3. CNC machine: Computer Numerical Control (CNC) machines are automated systems that utilize computer programming to control the movement and operation of cutting tools, allowing for highly accurate and repeatable processes. The core of CNC operation lies in the CNC controller, an embedded system that interprets the design specifications and converts them into a language, which the machine can understand. To operate a CNC machine, usually the process begins with the creation of a digital design using Computer-Aided Design (CAD) software. The designer then translates this design into a CNC compatible format, generating a G-code program. G-code is a series of alphanumeric commands that specify the machine's movements, tool changes, speeds, and other parameters. For example, a G01 command indicates linear motion, while G02 and G03 commands specify circular motions. The G-code program essentially serves as a set of instructions that guides the CNC machine through the entire machining process. Once the G-code program is ready, it is loaded into the control unit of CNC machine. The CNC controller interprets the G-code and sends signals to the motors and actuators, directing the movement of the cutting tool or the work piece. CNC machines typically operate on multiple axes, such as X, Y, and Z, allowing for precise control over the position of the tools in three-dimensional space. During operation, sensors and feedback mechanisms are often integrated into CNC machines to ensure accuracy. These may include encoders to monitor the position of each axis, probes to measure work piece dimensions, and other sensors for temperature or tool wear. This level of automation enhances efficiency, reduces human error, and facilitates the production of intricate components that would be challenging or impossible to manufacture manually.

The examples cited in this chapter are to inspire the students for building and creating robots using the knowledge acquired from the earlier chapters. This will guide the students to think, plan, design, and develop their projects following a systematic approach. Further, the course instructors can follow these examples to design new projects for the effective learning of the students engaged in solving real-world problems through their projects. More such projects can be explored at www.tezu.ernet.in/erl.

Index

If you have any concerns about our products,
you can contact us on
ProductSafety@springernature.com

In case Publisher is established outside the EU,
the EU authorized representative is:
Springer Nature Customer Service Center GmbH
Europaplatz 3, 69115 Heidelberg, Germany

Printed by Libri Plureos GmbH
in Hamburg, Germany